"*Reciprocal Landscapes* shows us what matters about landscape by revealing what matter is doing in it – where it came from, why it was taken, and how it was extracted, worked, fought over, and transported. Original in conception, rigorous in execution, Hutton's book is nothing less than a brilliant synthesis of materialisms 'historical' and 'new'; an incisive model for the critical analysis of landscape."

– Douglas Spencer, Director of Graduate Education and Associate Professor, Department of Architecture, Iowa State University

Reciprocal Landscapes

How are the far-away, invisible landscapes where materials come from related to the highly visible, urban landscapes where those same materials are installed? *Reciprocal Landscapes: Stories of Material Movements* traces five everyday landscape construction materials – fertilizer, stone, steel, trees, and wood – from seminal public landscapes in New York City, back to where they came from.

Drawing from archival documents, photographs, and field trips, the author brings these two separate landscapes – the material's source and the urban site where the material ended up – together, exploring themes of unequal ecological exchange, labor, and material flows. Each chapter follows a single material's movement: guano from Peru that landed in Central Park in the 1860s, granite from Maine that paved Broadway in the 1890s, structural steel from Pittsburgh that restructured Riverside Park in the 1930s, London plane street trees grown on Rikers Island by incarcerated workers that were planted on Seventh Avenue north of Central Park in the 1950s, and the popular tropical hardwood, ipe, from northern Brazil installed in the High Line in the 2000s.

Reciprocal Landscapes: Stories of Material Movements considers the social, political, and ecological entanglements of material practice, challenging readers to think of materials not as inert products but as continuous with land and the people that shape them, and to reimagine forms of construction in solidarity with people, other species, and landscapes elsewhere.

Jane Hutton is a landscape architect and teacher whose research looks at the expanded relations of material practice in design, examining linkages between the sources and construction sites of common building materials. Hutton is an Assistant Professor in the School of Architecture at the University of Waterloo.

Reciprocal Landscapes

Stories of Material Movements

JANE HUTTON

Routledge
Taylor & Francis Group

LONDON AND NEW YORK

First published 2020
by Routledge
2 Park Square, Milton Park, Abingdon, Oxon OX14 4RN

and by Routledge
52 Vanderbilt Avenue, New York, NY 10017

Routledge is an imprint of the Taylor & Francis Group, an informa business

© 2020 Jane Hutton

British Library Cataloguing-in-Publication Data
A catalogue record for this book is available from the British Library

Library of Congress Cataloging-in-Publication Data
Names: Hutton, Jane Elizabeth, 1976– author.
Title: Reciprocal landscapes : stories of material movement / Jane Hutton.
Description: Milton Park, Abingdon, Oxon ; New York, NY : Routledge, 2020. | Includes bibliographical references.
Identifiers: LCCN 2019018724 (print) | LCCN 2019022305 (ebook) | ISBN 9781138830684 (pbk) | ISBN 9781315737102 (ebk)
Subjects: LCSH: Landscape architecture—History. | Building materials—Transportation—History. | Landscaping industry.
Classification: LCC SB472.3 .H43 2020 (print) | LCC SB472.3 (ebook) | DDC 712—dc23
LC record available at https://lccn.loc.gov/2019018724
LC ebook record available at https://lccn.loc.gov/2019022305

ISBN: 978-1-138-83068-4 (pbk)
ISBN: 978-1-315-73710-2 (ebk)

Typeset in Sabon
by Apex CoVantage, LLC

For my parents: Bea & Peter, Bill & Bruce

This book contains excerpts and reworked material from several previously published articles including "Reciprocal Landscapes: Material Portraits in New York City and Elsewhere," *Journal of Landscape Architecture*, 2013, 8:1, pp. 40–47; "Inexhaustible Terrain" published in www.cca.qc.ca (Canadian Centre for Architecture, February 2017) CCA; "On Fertility: Night Soil, Street Sweepings, and Guano in Central Park," in the *Journal of Architectural Education*, 2014, 68:1, pp. 43–45; "Range of Motions: Granite Flow from Vinalhaven to New York City," in the *Harvard Design Magazine, Do you Read Me?*, No. 38, Spring – Summer, 2014, pp. 33–38; and "Trail of Stumps," in *Landscape Architecture Magazine*, May, 2013, pp. 116–126.

Contents

Figures

Note: Contemporary photographs are by author unless otherwise noted.

Introduction

Walking around a slate quarry in Bangor, Pennsylvania, the artist Robert Smithson gathered a bag of slate chips from the plunging, layered banks. Captured by the swirling patterns of brittle rock, he later described: "It was as though one was at the bottom of a petrified sea and gazing on countless stratigraphic horizons that had fallen into endless directions of steepness," and asked, "Yet if art is art it must have limits. How can one contain this 'oceanic' site?"[1] His response to that question would be a sculpture installed in 1968, a trapezoidal wood box – conjuring a one-point perspective – whose sharp lines contained the irregular slate shards within. Smithson described the sculpture as a three-dimensional map of the quarry, a fragment of an already fragmented slate landscape.

This piece, titled *Non-site (Slate from Bangor, Pa)*, is one of several *Non-sites* Smithson made that year. In each of them, he brought geological materials from quarries, mine dumps, and other industrial landscapes (which he called *sites*), and installed them as sculptures alongside maps and other site documentation into gallery spaces (or *non-sites*). Although they were sculptures set within gallery walls, they projected far beyond, to the site and back again – they were a means for dialectical thinking.[2] These pieces were more than fixed objects; they referred to the sites that they came from, and to their histories of geological and human transformation. The sites and non-sites were linked by material displacement from one to the other, but also through their differences. Sites were peripheral, overlooked spaces that supplied materials for urban development, while non-sites were central concentrations of cultural capital. Sites were real and physical, non-sites were abstract. Sites were the signified, non-sites the signifier.[3] Between the two was a space to think and see.[4]

And as the *Non-sites* were dialectical, so was landscape itself. In a 1973 essay titled "Frederick Law Olmsted and the Dialectical Landscape," Smithson examined the design and construction of Central Park in New York City. Olmsted and his partner, Calvert Vaux, were interested in the geological history of the urban parcel, and they made it visible in the park's design. By uncovering bedrock, carving tunnels of stone, and situating erratic boulders, the designers foregrounded the ever-changing landscape, referencing the park's connection to deep time and vast space. Olmsted understood the park, Smithson argued, not in isolation, but as a network of relations.[5] Just as the *Non-sites* sculptures

Figure 0.1 **Robert Smithson gathering slate in Bangor, Pennsylvania, 1968**

Source: © Holt-Smithson Foundation/SOCAN, Montréal/Licensed by VAGA at ARS, New York (2019). Unidentified photographer (Robert Smithson and Nancy Holt papers, 1905–1987, bulk 1952–1987. Archives of American Art, Smithsonian Institution).

Figure 0.2 **Robert Smithson,** *Non-site (Slate from Bangor, Pennsylvania)*, **1968**

Source: © Holt-Smithson Foundation/SOCAN, Montréal/Licensed by VAGA at ARS, New York (2019). Photograph by Walter Russell (Collection Dwan Gallery, Inc. In *Robert Smithson: Sculpture*, edited by Robert Hobbs. Ithaca: Cornell University Press, 1981, p. 125).

were not isolated objects, Smithson concluded the same of Olmsted's work: "A park can no longer be seen as 'a-thing-in-itself,' but rather as a process of ongoing relationships existing in a physical region."[6] Indeed, landscape – dynamically shaped by both human and nonhuman forces – was inherently suited for dialectical thinking, for thinking beyond itself.[7]

This book uses Smithson's *Non-site* as a structural prompt. *Reciprocal Landscapes: Stories of Material Movements* examines how the construction of a landscape in one place is related to transformation elsewhere.[8] In it, I explore five pairs of landscapes linked by the movement of a single material from one to the other. Each pair includes the landscape where it came from (analogous to Smithson's *site*) and the public landscape in which it was notably installed (analogous to his *non-site*). Smithson's terms inverse our expectations: he affirms the place of material production, and negates the place of material consumption. The materials featured in this book are everyday elements of constructed landscapes – fertilizer, stone, steel, trees, and wood – all installed in paradigmatic, designed landscapes in the borough of Manhattan in New York City. The five cases are arranged chronologically over the past century and a half, capturing changing material practices in relation to changing political and economic contexts. Source landscapes are diverse and far-flung, from coastal Peru to the American East Coast to northern Brazil. The paired landscapes are often quite distant from each other and seem only anecdotally connected – and yet they are intimately so.

These connections are the subject of this book.

By tugging on one seemingly thin thread, two sites are pulled together and a tangle of relations appears: how a particular geological or ecological condition became a desirable material commodity, how a material came to travel a certain route to the construction site, how the two sites (and the people and other species that occupy them) fared in this material transaction, and finally, how design and aesthetics participate in all of this. By circulating back and forth between of the two sites, it becomes difficult to see either in isolation.

Material Landscape Architecture

Materials – the soil mixes, seeds, aggregates, boards, blocks, and nursery stock – are the primary units of landscape construction, specified, acquired, and assembled into new forms. They range from abiotic to biotic, from raw to highly processed, from locally handcrafted to industrially produced and shipped around the globe. Contemporary design writing has pushed the discourse on materials beyond more conventional applications, foregrounding new and unusual materials and the ways in which designers use and misuse them, often within a particular site.[9] As designers increasingly address human-driven land and climate changes, they have started to focus on how materials perform beyond a project's fixed boundary: to filter water, extract contaminants from soil, support pollinators, or sequester carbon. Recent writing on landscape materials foregrounds their functional, hybrid biotic-abiotic, and technological nature, from Liat Margolis and Alexander Robinson's *Living Systems: Innovative Materials and Technologies for Landscape Architecture*[10] to the passive phyto-technologies discussed in Kate Kennon and Niall Kirkwood's *Phyto: Principles and Resources for Site Remediation*[11] to the broader environmental implications of specific materials in Meg Calkin's *Materials for Sustainable Sites*.[12] This book builds from these interests in the broader implications of material practice, and attempts to see these "externalities" as central to the field.

Formerly a task based on selecting locally available goods, material specification is now a global engagement. Manufacturers of construction materials, from lumber to metals to concrete, are among the world's largest corporations and offer an unlimited banquet of options, detached from locality. Everything is attainable for a price. Access to endless material options is tantalizing, but it also poses a series of fraught questions: Where did they come from? What are the social and ecological conditions of their manufacture? Who made them? What is left in their place? Regardless of how much you may want the backstory or to get materials from a specific location, information can be scarce and costs prohibitive. Everyone who specifies materials has absurd stories of how it was cheaper to get something from halfway across the globe than from next door. This is the reality of contemporary logistics, however illogical.

As one of the world's largest users of materials, water, and energy, as well as producers of considerable pollution and waste, the construction industry has been motivated by policy and economic interests to address the so-called externalities of building operations. Within the rubric of sustainable design and green building, the industry has developed various systems to account for the environmental impacts of construction. Lifecycle Analysis (LCA) is widely considered the most robust system for quantifying the lifecycle inputs, outputs, and environmental impacts of materials or small building components. However, it requires a significant amount of data and expertise, making it impractical for many typical construction projects.[13] More commonly known building rating systems – such as the U.S. Green Building Council's LEED (Leadership in Energy and Environmental Design) – examine an entire construction project rather than a single material.[14] They identify significant variables like water efficiency, energy consumption, waste production, and indoor environmental quality; provide a scale for scoring; and offer benchmarks for certification.[15] Other systems regulate and certify the chain of custody of specific materials, as is the case with the third-party monitoring of wood certified by the Forest Stewardship Council. While the sustainable building literature generally agrees that the pursuit of "sustainability" must involve environmental, social, and economic factors, most construction evaluation systems focus primarily, if not entirely, on the environmental.[16]

Modeled after LEED, the Sustainable Sites Initiative (SITES) – developed by the American Society of Landscape Architects and the Lady Bird Wildflower Center – specifically addresses landscape architecture projects. The program encourages designers to make decisions that strengthen "ecosystem services," a scheme to quantify and valuate the "services" to humans that ecosystems perform: the production of clear air, the regulation of erosion, pollination, and the support of human culture.[17] The guidelines encourage the adaptive reuse of existing materials, design for disassembly to minimize waste, the reuse of materials and plants, the specification of recycled content materials, the use of certified wood and local materials, and the reduction of emission-heavy chemicals. SITES also addresses plant nursery and material production, and credits are offered for supporting sustainable practices in both.[18] As with the building-focused certification systems, the intention is to foreground commonly externalized factors. These systems have successfully raised awareness and encouraged better practices; however, because their intention is to set generalizable guidelines, they are by necessity reductive. In contrast, this book offers little advice about "best" or "most sustainable" products to use, but rather, by diving into particular material histories, attempts to consider what material practice means on the ground, for people and other landscapes elsewhere.

Reciprocal Landscapes stems from a desire to think of construction materials not as fixed commodities or inert products, but as continuous with the landscapes they come from, and with the people that shape them. This perspective comes from my experience as part of white and non-white North American settler culture – part of a long, violent, and ongoing lineage of appropriation of indigenous lands for resource exploitation – and also my modern "privilege" of living in a service-based city that relies on other regional and international landscapes for sustenance. It is because of this that I find myself particularly separated from the world of production. And so my intention is to try to understand materials as fragments of other landscapes; as the livelihoods and habitats of people who live near them; as connections between the most tactile aspects of a design and the global circulation of matter driven by capitalism. All of these relations are consequential, and yet they can be hard to see. To attempt to see them, I use three approaches, each of which I'll elaborate on in this chapter: first, tracing a material's movements in order to see beyond its commodity status; second, comparing sites of material production and consumption in order to grapple with the ecologically unequal exchanges that accompany material transactions; and third, examining how these larger material relations play out in the designs and public engagement of designed landscapes.

Materials-in-Motion

While construction materials may appear to be fixed commodities, they are anything but fixed in time, space, or form. Materials change shape as they travel from geological deposit or forest to factory and design project to landfill, passing through human hands and tools. Think of all of the landscapes a material passes through or is physically contiguous with, the different forms it takes, and the people it interacts with along the way. Geographer David Harvey instructs us to do so by tracing our breakfast, and he follows his own:

> The coffee was from Costa Rica, the flour that made up the bread probably from Canada, the oranges in the marmalade came from Spain, those

in the Orange juice came from Morocco and the sugar came from Barbados. Then I think of all the things that went into making the production of those things possible – the machinery that came from West Germany, the fertilizer from the United States, the oil from Saudi Arabia.[19]

The exercise quickly reveals one's dependence upon yet obliviousness to the labor and environmental conditions of daily consumption.[20] Michael Pollan's book *The Omnivore's Dilemma: A Natural History of Four Meals*, which traces the ingredients of four different meals, has popularized the idea that these everyday products have much to tell us about contemporary North American society and its relationship to food, the people who produce it, and the land where it grows.[21]

That one can eat this food but not taste or know whether workers produced it in adequately waged or enslaved conditions is what Karl Marx famously called the fetishization of the commodity – the way in which markets conceal the source and labor conditions of its making, and the way in which the consumer of a good is systematically detached from its producer.[22] A commodity, as defined by Marx in *Capital*, is a valuable, useful, and exchangeable thing. It has *value* because human labor made it, it has *use-value* because someone wants to use it, and its *exchange-value*, or price, is set by the market. Exchangeable on the market, a commodity is equivalent to like-things from elsewhere: wheat is wheat, stone is stone, regardless of vastly different labor conditions or environmental consequences. So as land becomes raw material, and raw material becomes commodity, the rift between consumer and producer is ever widened.

This separation is almost too obvious to state as one rarely comes into contact with the person producing (or the land yielding) the goods that one consumes. If human labor and land are obscured through commodification, then tracing the lives of commodities can be an active method of uncovering, and also of de-essentializing, the commodity.[23] While it is easy to take something's commodity status for granted, as Arjun Appadurai argues in *The Social Lives of Things*, a commodity "is not one kind of thing rather than another, but one phase in the life of some things."[24] Only by observing "commodities-in-motion" – as they meander through commodity status, through various physical states, human relations, and locations – can we begin to understand them.[25]

Commodity chain analyses have typically focused on the economic exchanges between a series of firms, states, and consumers, but examining human labor and experience alongside a material trajectory can offer different insights; I'll cite three examples here. Geographer Elaine Hartwick argues that studying the social and symbolic dimensions of commodity chains brings together the otherwise separate experiences of consumers and producers. By being "geographical detectives" and uncovering the material relations of a product, Hartwick writes, one might better engage in meaningful activist praxis to alter these linkages, whether through advocacy campaigns or solidarity movements with workers elsewhere.[26] In a second example, in her book *In the Aura of a Hole: Exploring Sites of Material Extraction*, artist Laurie

Palmer travels to extraction landscapes of eighteen elements, from iron to lead to copper, delving into the physical contexts and narratives of people living and working near them.[27] Palmer emphasizes how commodities are embedded in a specific place, and shows us their complicated histories and long-standing human witnesses. Finally, focusing on the later end of the construction material trajectory – the assembly of buildings – the project *Who Builds Your Architecture?* (WBYA?) asks architects to address the working conditions of people constructing the buildings they design. Just as material production has become a global enterprise, so too has architecture; designers today design for places far from their familiar contexts, with little connection to the labor conditions in those places.[28] WBYA? asks architects to consider the ethical and political questions that this raises, connecting designers with laborers elsewhere.

If the act of tracing materials and commodities leads to people, it also inevitably leads to land. Tracing wood, stone, iron, and polymers leads to forest communities, sedimentary deposits, iron-rich seams, and fossil fuel beds. Tracing materials back to the land can reveal how certain properties (the durability of certain wood or the shininess of a stone, for example) are not merely "useful" attributes, but how they are physically related to unique, local biophysical conditions. It is easy to see landscapes as "natural resources," standing reserves of materials ready for the taking, but they are of course more than this. They are complex ecosystems that support many interconnected beings; they are the physical basis of human sustenance and local culture.

At the hinge point between land and commodity, materials teeter uncomfortably between that which is considered natural and that which is not, between that which is intrinsically valuable and between that which is valuable for human use. By focusing on the moment between land and commodity, when a tree is a tree and when it becomes fungible, saleable lumber, or when a particular geological deposit becomes a valuable ore, we can witness this continuity and un-see the commodity for just a moment. Compared to materials used in buildings, materials used in landscape architecture are often less processed or closer to a "raw" form. An eastern white cedar 2×4 and a New York bluestone slab, for example, are recognizably related to the landscapes they come from, and their product names even suggest their geographical source. In comparison to architecture, within landscape architecture, the existing site conditions – soil, vegetation, and contamination – are also "materials" that are inevitably incorporated into the project. To a casual observer, a simple tree planted in a grassy surface may appear completely "natural," but in fact may be a highly manufactured complex including specially bred plant material, chemical fertilizers and additives, engineered soils, and polymer turf.

Because landscape materials are neither clearly natural nor human-made, thinking about them can disrupt unhelpful binaries. Materials shape-shift as they move in and out of human controlled systems, challenging us to think of them as both formed through human action and also as having lives of their own. If materials are not only for human use, how might we consider them outside of a purely instrumental light? If we could see matter as having agency, how might this affect the way we work with it, build with it, and live with it?

Reacting to the post-modern tendency to privilege semiotic readings of the world over physical ones and to separate meaning from matter, materialist thinkers – from feminist and queer studies to science and technology studies to political science – have offered a range of modes for thinking through these questions.[29] Matter, as feminist theorist Karan Barad argues, "is not little bits of nature, or a blank slate, surface or site passively awaiting signification, . . . immutable or passive."[30] Instead, because matter intra-acts within the world, it is inherently agentic, discursive, and an important participant in the making of the world.[31]

While we typically focus on the ways in which humans control, shape, and transform matter, the physical qualities of matter are powerful in their own right, reciprocally shaping human activity. In studying extractive industries in the Amazonian basin, geographers Stephen Bunker and Paul Ciccantell remind us of how the entire industrial apparatus of technology, commodity, and market is constrained by the physical qualities of the materials themselves.[32] It is the heaviness of certain ores, for example, which truly drive industrial location, and a material's exploitation will be determined not only by its useful properties, but also its non-useful ones. Geographers Karen Bakker and Gavin Bridge urge us to pay attention to all of the ways that materials don't "cooperate," and how their unpredictability or unruliness might disrupt or irritate capital accumulation.[33] Why consider such nuance? To do so challenges understandings of materials as inert and wholly subservient to human agenda; this alone is an important starting point for seeing materials (and the more-than-human world that they comprise) as more than stuff to use.

Material Exchange

If tracing material trajectories reveals how they are connected to land and people, the second approach of *Reciprocal Landscapes* pairs and compares two of these landscapes: a material "origin" landscape and a designed landscape in which the material was installed. The notion of an origin is suspicious from the start: where does a material begin anyways? By looking at these two sites together, we can observe how single transactions produce radically different conditions. It is possible to see each landscape through the other, and perhaps, grapple with the baffling whole.

At the scale of the planet, humans metabolize matter for construction and agriculture at a rate ten times that of global geological processes alone.[34] Humans have always reshaped their environments; however, as environmental historian J. R. McNeill has meticulously documented, due to spiking population, rising per capita consumption, the mass production of consumer goods, infrastructure, urbanization, and war during the twentieth century, this anthropogenic transformation accelerated like never before.[35] For the first time, the magnitude of *local* human activities produced new *global* conditions in the lithosphere, biosphere, hydrosphere, and atmosphere of the earth. And in the years spanned in this book, the global material flow began to surge.

A graph of the global consumption of materials over the twentieth century, compiled by Fridolin Krausmann and colleagues, shows a line lunging to the upper right-hand corner.[36] Over those hundred years, the earth's social and industrial metabolism – the inputs and wastes associated with human-driven developments – exploded: global material consumption multiplied eight-fold, skyrocketing in the post-war period, now around 60 gigatons of material per year. Not only did the *scale* of material use change, its *composition* changed as well. While at the beginning of the twentieth century humans consumed primarily biomass materials (crops, fodder, and wood), as the century progressed there was a switch to minerals, reflecting a shift from agrarian to industrial economies. By the end of the century, humans extracted and used thirty-four times as many construction minerals (cement, asphalt, sand, and gravel), and twenty-seven times as many metal and industrial minerals (iron, copper, aluminum, etc.), as they had in 1900.[37] This change in consumption signaled a material paradigm shift: from organic to mineral, from renewables to finite resources, and from materials that move quickly through society (like biomass and combusted fuel) to those which accumulate and reside in a place (like metals and concrete in urban infrastructure).[38] As urban areas expanded, infrastructure multiplied, per-capita material consumption skyrocketed, and wars raged, these materials – mostly sand, aggregates, cement, and metals – migrated and accumulated in new strata around the planet.

Krausmann's graph illustrates the coupling of material extraction and capital accumulation. As GDP soared, so too did material and energy consumption. The only lulls in the world's upward consumption occurred during periods of economic stagnation, during the global economic crisis in the 1930s, World Wars I and II, and the oil crisis of the 1970s.[39] And while some improvement of material efficiency occurred, these efficiencies, or "dematerializations," never led to a reduction in consumption; on the contrary, ever-new mechanisms for consumption emerged.[40] This tight pairing between materials and capital suggests something profound about the history of capitalism; as Jason W. Moore puts it: "Natures were appropriated. Capital was accumulated. Wastes were dumped overboard."[41] The material exploitation shown in Krausmann's graph, and associated environmental changes, are not a *consequence* of capitalism at work, but rather capitalism's ecological *modus operandi*.[42] Capitalism, Moore argues, is a system predicated on the creation of what he calls "cheap natures": cheap labor, food, energy, and raw materials; when these cheap materials are exhausted in one place, capitalism moves to the next. The cases in this book reflect these patterns, from soil exhaustion to resource eradication and deforestation. And likewise, the anthropogenic ecological crises that we witness today are not the system malfunctioning; they reflect instead the logical outcomes of capital accumulation based on using things up as if there were no limits or costs.[43]

This global material flow plays out in highly differentiated ways on the ground: deeper quarries in one place, material improvements in another; toxic deposition in one place, refined minerals put to use in another. Acknowledging that certain places and people (typically capitalist core nations or cities) gain resources and benefits at the expense of others (typically in the so-called periphery) is what Alf Hornborg calls the "zero-sum-world" perspective.[44] Research into how the use of materials (or "resources") is structurally inequitable and geographically distributed has identified linkages between material flow analysis, ecological conflicts, and the notion of ecologically unequal exchange.[45] Looking along the material chain from extraction to production, one can observe some broad, basic tendencies that point to how one instance of material exchange can produce extremely different realities in different places. First, finished products are priced disproportionately higher than the raw materials used to produce them, which incentivizes more material processing[46]; second, local extractive economies (often in poorer areas) tend to decrease in power over time as available ores become less pure, scarcer, and more remote. These economies' wealth is based on rates of natural production, typically slower than desired, and so to extract more is to over-harvest and degrade one's own resources[47]; third, core industrial economies that process and finish material products tend to increase in power over time: they acquire cheaper and greater access to raw materials, develop technology and infrastructure to increase extraction and transportation, and develop financial instruments and state collaborations to then acquire even cheaper materials in still greater quantities.[48] And finally, core countries tend to export their polluting industries to poorer countries.[49]

In this book, the examination of paired sites – a production site and a material use site – is a simplified conceit, but it stands to consider how different sites benefit from or bear the risks of material exchange. In these cases, the extraction or production sites reflect a range of conditions from developing global economies to other American cities. The material use sites are all located in a financial and cultural epicenter of a capitalist core industrial nation, New York City, and more specifically, in the borough of Manhattan. Beyond the physical movement of a material from production to consumption sites, there are many other types of exchange. For example, banks and institutions in New York City are involved in financing, insuring, and facilitating transportation systems and land purchases far from the city limits – indeed, across the globe.

Material Culture

How do these larger material movements and exchanges relate to the more typical concerns of landscape architecture, such as how materials should be assembled within a certain place, what meanings are conveyed through their form, and how the public engages with different projects? Landscape historian Marc Treib argues that meaning is accrued in many ways over time by diverse populations and in ways that continue to change; the meanings that a designer might intend are in fact mutable, reflections of the time and public opinion.[50]

As historian of science Lorraine Daston writes, "things talk," and for the purposes of this book, landscapes talk as well; they convey ideas and cultural meanings both through the form that they take, but also through how they are made and their material stuff.[51]

Designed landscapes *mean* many things to many people, but we can see them as models of human-nature relations: models of being with, working with, and interacting with all that is more-than-human.[52] They have long-held didactic and representational roles to teach, inform, and develop ideas about honoring, cultivating, and changing land. From the classical gardens of the Middle East, Asia, and Europe depicting stories of creation and paradise, to North American urban commons depicting enclosures and colonial power, to contemporary post-industrial projects prefiguring new relationships with wild species, designed landscapes reflect cultural attitudes about the control and change of land. Landscapes are models *in situ*, as well as experimental testing grounds for new forms of changing land elsewhere.[53]

And so while designed landscapes model ideas about human relationships to land, they also engage in *real* ones through their material making. By looking at the cultural intentions behind a landscape design, we can begin to consider the following: If designed landscapes seek to reflect harmonious human-nature relationships, how well do these align with the ways that landscapes are actually made? How do the conceptual agendas of designed landscapes engage, ignore, or contradict the violent and problematic aspects of their material production? How do designers make decisions about material sourcing, and how are these decisions constrained by market availability and professional standards and norms? While most of these questions go beyond the scope of this book, the cases point to relationships between contemporaneous environmental thinking, material availabilities, and how the public perceives, adopts, or resists certain forms of making.

Reciprocal Landscapes

All of the cases in this book have one foot in Manhattan and one foot somewhere else. Facing Manhattan from the Brooklyn Bridge on the East River, you can see how the geology of the island has prefigured the city's famous skyline: an eruption of towers downtown gives way to another mass mid-town, where the bedrock Manhattan schist is close enough to the surface to support the skyscrapers that would come to be symbols of modernity. While it is hard not to focus on each of the iconic buildings, it is fun to try to imagine the city instead as a multi-millennial material flow. You could see waves of geologic and ecological transformation, and then zooming in on the last thousand years, wave upon wave of human occupation, from Lenape inhabitants to coinciding waves of immigrants and colonizers, each laying a different material mixture as new strata of earth, wood, stone, steel, concrete, and glass.

Emerging in the late nineteenth and early twentieth centuries as a global financial center, New York City reflects an intensified example of urbanization

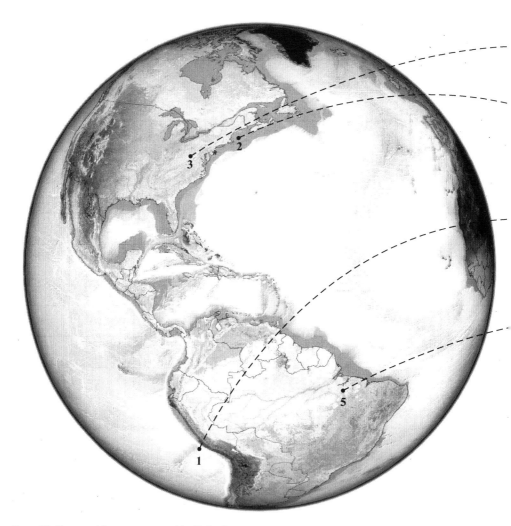

Figure 0.3 **Five material movements traced in this book:**

1 Guano from the Chincha Islands, Peru, to Central Park
2 Granite from Vinalhaven, Maine, to Broadway
3 Steel from Pittsburgh to Riverside Park
4 Trees from Rikers Island to Seventh Avenue
5 Tropical hardwood from northern Brazil to the Highline
* Location of designed landscapes within Manhattan, shown shaded

Source: Base maps © 2019 Google.

4

* ⊢ 1 mi ⊣

on the East Coast of the United States. The city has inspired a rich literature examining the production of space and, in particular, non-human and bio-physical contexts.[54] Geographer Matthew Gandy's book *Concrete and Clay* explores the development of New York City through the very hydrological and ecological systems in which it has been built. Gandy's environmental history reveals how nature has been "re-worked" both as physical material and to serve political ideologies and means of environmental management.[55] In his book *Manhattan Atmospheres*, David Gissen looks at the environmental and economic crises of the late twentieth century, exploring how architects responded by bringing nature inside to a more privatized realm.[56] Their work challenges more typical narratives about the development of New York City; the land is not a passive background to human achievement but rather a key actor in the historical transformation of the metropolis.

The five cases in this book draw from the early 1860s to the present. The landscapes where the materials ended up range from canonical design precedents (Central Park in the late nineteenth century to Riverside Park in the early twentieth century to the High Line in the twenty-first) to infrastructural achievements (the successive movements to repave and replant city streets in the nineteenth and twentieth centuries). Together, they make up a loose history of predominant construction materials, technology, and attitudes about resource exploitation and labor. The changing political-economic contexts, from industrial capitalism to Fordist Keynesianism to neoliberalism, provide a framework for understanding each material case – alongside changing urban, social, and environmental conditions, as well as material practices.

Each chapter (and its material) also raises a particular theme. The first chapter traces a small quantity of guano fertilizer used to enrich the lawns of Central Park in the 1860s back to the Chincha Islands of coastal Peru. As powerful urbanizing centers like New York City expanded in the mid-nineteenth century, so too did the scale of their peripheral agricultural lands, and their appetite for high-potency, foreign fertilizers. Landscape architect Frederick Law Olmsted experimented with the new fertilizer (alongside other composted wastes) in Central Park, designed as a microcosm of the agricultural experience for urban populations alienated from the countryside. As the global trade in mineral fertilizers supercharged the industrialization of modern agriculture, guano deposits on the Chincha Islands were quickly exhausted, extracted by enslaved Chinese workers under horrific conditions. The case of guano illustrates the growing metabolic rift of the nineteenth century, anxieties about disconnections from local organic cycles, and interwoven narratives of the exhaustion of both soil and enslaved workers.

As the city became a global financial center in the late nineteenth century, crowded streets and broken pavements interfered with the smooth flow of capital and goods. Chapter 2 follows granite paving blocks from the Fox Islands of coastal Maine to the smoothened pavement of Broadway in the 1890s. Cities along the Atlantic seaboard grew rapidly at this time, with paved street grids and large institutional buildings, bridges, and monuments. Granite, hard and noble, was a material of choice. Large government contracts profited quarry owners and drew granite cutters to small towns like Vinalhaven in the Fox Islands. As the islands' quarries deepened and New York City's streets hardened, quarry owners profited at the expense of precarious and poorly compensated workers. In response, Maine quarry workers organized the country's first stone-cutting union. The heavy flow of granite from Maine to New York was disrupted along the material trajectory as paving workers in New York went on strike in solidarity with Maine workers. The chapter explores how material flow, rather than an abstract vector, is carried by human hands, and can be powerfully disrupted by them.

With the city's vertical expansion in the early twentieth century, granite gave way to concrete reinforced with steel. The third chapter follows structural steel from one riverside to another in the 1930s: from United States Steel Corporation's Carrie Blast Furnaces on the Monongahela River near Pittsburgh to Riverside Park on the Hudson River in Manhattan. In the early 1900s, the steel industry exemplified the strategies of monopoly capitalism by capturing, integrating, and reorganizing massive holdings of ore deposits, infrastructure, and labor across the country. At the same time, steel materially reorganized the modern city – spanning infrastructure, structuring concrete, and supporting towers – but also immaterially through speculative real estate development in Manhattan. New Deal legislation stimulated a lagging steel industry through massive steel infrastructure investments (in places like New York City), but also new labor legislation that enabled the struggle towards collective bargaining (centered in Pittsburgh). This chapter looks at the role that steel and progressive legislation played in the construction of the new modern public landscape, and in the support of organized industrial labor. It also examines the inherent tensions between industrial development and the environmental risks born by residents – and the ways in which landscape architects mediate between them.

By the late 1950s, the widespread erosion of progressive social programs, the shift from more stable Fordist factory work to more precarious service work, and ongoing racial segregation manifested in a decaying public realm and the uneven provision of municipal services. Chapter 4 follows juvenile London plane trees (*Platanus x acerifolia*) that grew in a Parks Department tree nursery on Rikers Island – tended by incarcerated workers – to a community-driven tree planting on Seventh Avenue in Harlem. From the city's point of view, street trees, with their grand canopies and health benefits, offered a quality of urban living that might bring fleeing urbanites back to the city. In Harlem, communities of color had long fought for decent municipal services, housing justice, healthcare provision, and community amenities – including the improvement of the streetscape with trees. At the time, these trees (and tens of thousands more being planted throughout the city) came from Rikers Island. On Rikers Island, the city developed a tree nursery that could provide a cheap and continuous supply of trees for streets and municipal parks, using the unpaid labor of incarcerated people. This chapter looks at the street tree as an indicator of urban health, a marker of unequally distributed environmental amenities, and a character of gentrification as public space was increasingly privatized in the late twentieth century. The street tree offers a reflection on the degree of continual care required for both humans and trees to thrive in urban habitats.

Finally, the fifth chapter links the Amazon rainforest of northern Brazil with the High Line in Manhattan by following the movement of ipe (*Tabebuia spp.*), multiple slow-growing deciduous tree species whose wood is extremely durable, hard, and rot-resistant. Because the harvest of ipe targets sparsely distributed, old-growth trees, and participates in patterns of large-scale deforestation driven by global markets, its use is controversial. At the same time, the use of durable materials is a central tenet of sustainable building best practices. The chapter examines the theme of durability and obsolescence, through the material of tropical hardwood, but also through the re-used elevated rail structure in Chelsea (the same structure built in the 1930s as part of the West Side Improvement, which included the renovation of Riverside Park). The case of ipe lumber highlights contemporary conditions of globalized neoliberalism, linking new post-industrial landscapes of leisure with the migration of industry elsewhere, the increased access to global lumber, and the outsourcing of ecological risk.

If the process of material exchange is often deeply unequal, why use the term "reciprocal" in the title of this book? Reciprocity is a robotic-sounding word with a beautiful meaning: it connotes a gift exchange, a give-and-take, a relationship of mutual benefit. Contemporary urbanization through globalized capitalism is anything but reciprocal. Alf Hornborg writes that one of the major illusions of sustainability discourse is the assumption that market prices indicate a reciprocal relationship.[57] Botanist and Potawatomi writer Robin Wall Kimmerer writes in depth on the concept of reciprocity, of the mutual interdependencies that she observes as a plant scientist and student of indigenous knowledge. Kimmerer likens reciprocity to gratitude: that to be alive, to build, to eat, or to make art, is to exchange with others and to recognize this is to be thankful. She writes of how the traditional Haudenosaunee Thanksgiving Address, repeated for generations, is a refrain for this gratitude and also establishes a political document and social contract:

> Cultures of gratitude must also be cultures of reciprocity. Each person, human or no, is bound to every other in a reciprocal relationship. Just as all beings have a duty to me, I have a duty to them. If an animal gives its life to feed me, I am in turn bound to support its life. If I receive a stream's gift of pure water, then I am responsible for returning a gift in kind.[58]

"If we want to grow good citizens," Kimmerer continues, "then let us teach reciprocity."[59] In this book, *Reciprocal Landscapes*, the word reciprocal is not intended to soften, conceal, or suggest equanimity between the sites of production and consumption. Rather, "reciprocal" is used in an aspirational sense. When joined with "landscape," it suggests the inextricable interdependencies that humans share with the more-than-human world, that consumers share with producers, and that all beings and things share with each other. By tracing each material, examining the unequal dynamics of exchange along its path, and probing a project's ideological agendas alongside these material relations, these cases offer a set of points from which more lines can be drawn. Moving back and forth between the distant production site and designed landscape, the aim is to bring them, at least conceptually, closer together. *Reciprocal Landscapes* is above all a thought experiment: What if we looked at materials not simply as single-purpose products or commodities, but instead as continually changing matter that takes different forms, and is shaped by – but also shapes – others? And more broadly, how might understanding these so-called externalities of development inflect new forms of material practice in solidarity with people, other species, and landscapes elsewhere?

Notes

1 Robert Smithson, "A Sedimentation the Mind: Earth Projects," *Art Forum*, September 1968, republished in *Robert Smithson: The Collected Writings*, ed. Jack Flam (Berkeley: University of California Press, 1996), 110–11.

2 Robert Smithson, "A Provisional Theory of Non-Sites (1968)," in *Robert Smithson: The Collected Writings*, ed. Jack Flam (Berkeley: University of California Press, 1996), 364.

3 Lawrence Alloway, "Sites / Nonsites," in *Robert Smithson: Sculpture*, ed. Robert Hobbs (Ithaca: Smithmark Pub, 1983), 42–3.

4 Robert Smithson, "A Provisional Theory of Non-Sites (1968)."

5 Robert Smithson, "Frederick Law Olmsted and the Dialectical Landscape," in *Robert Smithson: The Collected Writings*, ed. Jack Flam (Berkeley: University of California Press, 1996), 160; See Elizabeth Meyer, "Site Citations: The Grounds of Modern Landscape Architecture," in *Site Matters: Design Concepts, Histories, and Strategies*, ed. Carol J. Burns and Andrea Kahn (New York: Routledge, 2005), 98.

6 Here, Smithson challenges Immanuel Kant's notion of "thing-in-itself" (*ding an sich*), or a thing independent of human observation, as a formal ideal, and instead emphasizes things as manifolds of relations. Thanks to Alex Livingston for this reference.

7 Like Robert Smithson's assertion about Frederick Law Olmsted's Central Park and landscape writ large, Cultural Geographer Don Mitchell has suggested that landscape is an inherently dialectical subject, see Don D.M. Mitchell, "Cultural Landscapes: The Dialectical Landscape – Recent Landscape Research in Human Geography," *Progress in Human Geography* 26, no. 3 (June 1, 2002): 381–9.

8 "Reciprocal" is a loaded term that I will address later on.

9 See, for example Blaine Brownell, *Transmaterial: A Catalog of Materials that Redefine Our Physical Environment* (New York: Princeton Architectural Press, 2006), and Blaine Brownell, *Material Strategies: Innovative Applications in Architecture* (New York: Princeton Architectural Press, 2013).

10 Liat Margolis and Alexander Robinson, *Living Systems: Innovative Materials and Technologies for Landscape Architecture* (Basel and Boston: Birkhäuser, 2007).

11 Kate Kennen and Niall Kirkwood, *Phyto: Principles and Resources for Site Remediation and Landscape Design* (New York: Routledge, 2015).

12 Meg Calkins, *Materials for Sustainable Sites: A Complete Guide to the Evaluation, Selection, and Use of Sustainable Construction Materials* (Hoboken: Wiley, 2008). See also: William Thompson and Kim Sorvig, *Sustainable Landscape Construction: A Guide to Green Building Outdoors* (Washington, DC: Island Press, 2000).

13 Arpad Horvath, "Construction Materials and the Environment," *Annual Review of Environment and Resources* 29 (2004): 188.

14 See also, BREEAM (Building Research Establishment Environmental Assessment Method) in the UK.

15 Ghassan Marjaba and Samir Chidiac, "Sustainability and Resiliency Metrics for Buildings: A Critical Review," *Building and Environment* 101 (May 2016): 116–25.

16 Ibid. For a recent study looking at the integration of social sustainability measures in an LCA process, see Uzzal Hossain, Chi Sun Poon, Ya Hong Dong, Irene M.C. Lo, and Jack C.P. Cheng, "Development of Social Sustainability Assessment Method and a Comparative Case Study on Assessing Recycled Construction Materials," *The International Journal of Life Cycle Assessment* 23, no. 8 (August 2018): 1654–74.

17 Meg Calkins, *The Sustainable Sites Handbook: A Complete Guide to the Principles, Strategies, and Best Practices for Sustainable Landscapes* (Hoboken: Wiley, 2012).

18 Sustainable Sites Initiative, "Guidelines and Performance Benchmarks, 2009," pp. 124–38, accessed October 12, 2018, https://digital.library.unt.edu/ark:/67531/metadc31157/m2/1/high_res_d/Guidelines%20and%20Performance%20Benchmarks_2009.pdf.

19 David Harvey, "Editorial: A Breakfast Vision," *Geographical Review* 3, no. 1 (1989).

20 David Harvey, "Between Space and Time: Reflections on the Geographical Imagination," *Annals of the Association of American Geography* 80, no. 3 (1990): 422.

21 Michael Pollan, *The Omnivore's Dilemma: A Natural History of Four Meals* (New York: Penguin, 2007).

22 Karl Marx and Ernest Mandel, *Capital: A Critique of Political Economy*, trans. Ben Fowkes (New York: Penguin, 2004), 164–5.

23 Noel Castree, "The Geographical Lives of Commodities: Problems of Analysis and Critique," *Social & Cultural Geography* 5, no. 1 (March 2004): 23.

24 Arjun Appadurai, ed., *The Social Life of Things: Commodities in Cultural Perspective* (Cambridge: Cambridge University Press, 1988), 17.

25 Ibid., 16. Within geography, an extensive literature focuses on following the "lives," "chains," "circuits," and "networks" of commodities, in order to better understand how power plays out along these lines and challenge the idea that commodities are fixed, given entities. See, for example, Peter Jackson, "Commercial Cultures: Transcending the Cultural and the Economic," *Progress in Human Geography* 26, no. 1 (2002): 3–18.

26 Elaine R. Hartwick, "Towards a Geographical Politics of Consumption," *Environment and Planning* 32 (2000): 1177–92.

27 Laurie Palmer, *In the Aura of a Hole: Exploring Sites of Material Extraction* (London: Black Dog Publishing, 2015).

28 "Who Builds Your Architecture?: A Critical Field Guide," 2017. "WBYA_Guidebook_spreads.Pdf," accessed October 21, 2018, http://whobuilds.org/wp-content/uploads/2017/02/WBYA_Guidebook_spreads.pdf.

29 See for example, Jane Bennett, *Vibrant Matter: A Political Ecology of Things* (Durham: Duke University Press, 2009), Peter Jackson, "Rematerializing Social and Cultural Geography," *Social & Cultural Geography* 1, no. 1 (September 1, 2000): 9–14; Sarah Whatmore, "Materialist Returns: Practicing Cultural Geography in and for a More-Than-Human World," *Cultural Geographies* 13 (2006): 600–9.

30 Karen Barad, "Posthumanist Performativity: Toward an Understanding of How Matter Comes to Matter," *Signs: Journal of Women in Culture & Society* 28, no. 3 (Spring 2003): 801.

31 Ibid., 821–3. Geographer Juanita Sundberg reminds of how post-human theory has also had a tendency to reproduce colonial relationships, by asserting totalizing narratives, See Juanita Sundberg, "Decolonizing Posthumanist Geographies," *Cultural Geographies* 21, no. 1 (2004): 33–47.

32 Stephen G. Bunker and Paul S. Ciccantell, "Globalizing Economies of Scale in the Sequence of Amazonian Extractive Systems," in *Globalization and the Race for Resources* (Baltimore: JHU Press, 2005), 33.

33 Karen Bakker and Gavin Bridge, "Material Worlds? Resource Geographies and the 'Matter of Nature'," *Progress in Human Geography* 30, no. 1 (February 1, 2006): 18.

34 Bruce H. Wilkinson, "Humans as Geologic Agents: A Deep-time Perspective," *Geology* 33, no. 3 (2005): 161–4, 161.

35 J.R. McNeill, *Something New Under the Sun: An Environmental History of the Twentieth Century World* (New York: WW Norton, 2001).

36 Fridolin Krausmann, Simone Gingrich, Nina Eisenmenger, Karl-Heinz Erb, Helmut Haberl, and Marina Fischer-Kowalski, "Growth in Global Materials Use, GDP and Population during the 20th Century," *Ecological Economics* 68, no. 10 (August 15, 2009): 2699.

37 Ibid.

38 Ibid., 2701.

39 Ibid., 2702.

40 Ibid.

41 Jason W. Moore, *Capitalism in the Web of Life: Ecology and the Accumulation of Capital* (New York: Verso, 2015), 291.

42 Ibid., 111–40.

43 Ibid., 291–305.

44 Alf Hornborg, "Zero-Sum World: Challenges in Conceptualizing Environmental Load Displacement and Ecologically Unequal Exchange in the World-System," *International Journal of Comparative Sociology* 50, no. 3–4 (June 1, 2009): 237–62.

45 See Joan Martinez-Alier, *The Environmentalism of the Poor: A Study of Ecological Conflicts and Valuation* (Cheltenham: Edward Elgar, 2002); Alf Hornborg, "Towards an Ecological Theory of Unequal Exchange: Articulating World System Theory and Ecological Economics," *Ecological Economics* 25, no. 1 (April 1998): 127–36.

46 Hornborg, "Towards an Ecological Theory of Unequal Exchange".

47 Bunker and Ciccantell, "Globalizing Economies of Scale in the Sequence of Amazonian Extractive Systems," 225.

48 Ibid., 224.

49 Roldan Muradian and Stefan Giljum, "Physical Trade Flows of Pollution-Intensive Products: Historical Trends in Europe and the World", in *Rethinking Environmental History: World-system History and Global Environmental Change*, eds. A. Hornborg, J.R. McNeill and J. Martinez-Alier (Walnut Creek, CA: AltaMira Press, 2007), 307–25.

50 Marc Treib, "Must Landscapes Mean?: Approaches to Significance in Recent Landscape Architecture," *Landscape Journal* 14 (January 1, 1995): 46–62.

51 Lorraine Daston, ed., *Things That Talk: Object Lessons from Art and Science* (New York: Zone Books, 2007), 16–17.

52 See Jane Hutton, ed., *Landscript 5: Material Culture: Assembling and Disassembling Landscapes* (Berlin: Jovis, 2018).

53 Not only do designed landscapes reference power, they have often been explicitly designed to display and expand it. See for example, Chandra Mukerji, "Space and Political Pedagogy at the Gardens of Versailles," *Public Culture* 24, no. 3 (2012): 509–34.

54 See for example, Eric W. Sanderson, *Mannahatta: A Natural History of New York City* (New York: Harry N. Abrams, 2013); Ted Steinberg, *Gotham Unbound: The Ecological History of Greater New York* (New York: Simon & Schuster, 2015).

55 Matthew Gandy, *Concrete and Clay: Reworking Nature in New York City* (Cambridge, MA: The MIT Press, 2003).

56 David Gissen, *Manhattan Atmospheres: Architecture, the Interior Environment, and Urban Crisis* (Minneapolis: University of Minnesota Press, 2014).

57 Hornborg, "Zero-Sum World," 256.

58 Robin Wall Kimmerer, *Braiding Sweetgrass: Indigenous Wisdom, Scientific Knowledge and the Teachings of Plants* (Minneapolis: Milkweed Editions, 2015), 115.

59 Ibid., 116.

Bibliography

Alloway, Lawrence. "Sites/Nonsites." In *Robert Smithson: Sculpture*, edited by Robert Hobbs, 41–5. Ithaca: Smithmark Pub, 1983.

Appadurai, Arjun, ed. *The Social Life of Things: Commodities in Cultural Perspective.* Cambridge: Cambridge University Press, 1988.

Bakker, Karen, and Gavin Bridge. "Material Worlds? Resource Geographies and the 'Matter of Nature'." *Progress in Human Geography* 30, no. 1 (February 1, 2006): 5–27.

Barad, Karen. "Posthumanist Performativity: Toward an Understanding of How Matter Comes to Matter." *Signs: Journal of Women in Culture & Society* 28, no. 3 (Spring 2003): 801–31.

Bennett, Jane. *Vibrant Matter: A Political Ecology of Things.* Durham: Duke University Press, 2009.

Brownell, Blaine. *Material Strategies: Innovative Applications in Architecture.* New York: Princeton Architectural Press, 2013.

Brownell, Blaine. *Transmaterial: A Catalog of Materials That Redefine Our Physical Environment.* New York: Princeton Architectural Press, 2006.

Bunker, Stephen G., and Paul S. Ciccantell. "Globalizing Economies of Scale in the Sequence of Amazonian Extractive Systems." In *Globalization and the Race for Resources.* Baltimore: JHU Press, 2005.

Calkins, Meg. *Materials for Sustainable Sites: A Complete Guide to the Evaluation, Selection, and Use of Sustainable Construction Materials.* Hoboken: Wiley, 2008.

Calkins, Meg. *The Sustainable Sites Handbook: A Complete Guide to the Principles, Strategies, and Best Practices for Sustainable Landscapes.* Hoboken: Wiley, 2012.

Castree, Noel. "The Geographical Lives of Commodities: Problems of Analysis and Critique." *Social & Cultural Geography* 5, no. 1 (March 2004): 21–35.

Daston, Lorraine, ed. *Things that Talk: Object Lessons From Art and Science.* New York: Zone Books, 2007.

Gandy, Matthew. *Concrete and Clay: Reworking Nature in New York City.* Cambridge, MA: The MIT Press, 2003.

Gissen, David. *Manhattan Atmospheres: Architecture, the Interior Environment, and Urban Crisis.* Minneapolis: University of Minnesota Press, 2014.

Hartwick, Elaine R. "Towards a Geographical Politics of Consumption." *Environment and Planning* 32 (2000): 1177–92.

Harvey, David. "Between Space and Time: Reflections on the Geographical Imagination." *Annals of the Association of American Geography* 80, no. 3 (1990): 418–34.

Harvey, David. "Editorial: A Breakfast Vision." *Geographical Review* 3, no. 1 (1989).

Hornborg, Alf. "Towards an Ecological Theory of Unequal Exchange: Articulating World·System Theory and Ecological Economics." *Ecological Economics* 25, no. 1 (April 1998): 127–36.

Hornborg, Alf. "Zero-Sum World: Challenges in Conceptualizing Environmental Load Displacement and Ecologically Unequal Exchange in the World-System." *International Journal of Comparative Sociology* 50, no. 3–4 (June 1, 2009): 237–62.

Horvath, Arpad. "Construction Materials and the Environment." *Annual Review of Environment and Resources* 29 (2004): 181–204.

Hossain, Uzzal, Chi Sun Poon, Ya Hong Dong, Irene M.C. Lo, and Jack C.P. Cheng, "Development of Social Sustainability Assessment Method and a Comparative Case Study on Assessing Recycled Construction Materials." *The International Journal of Life Cycle Assessment* 23, no. 8 (August 2018): 1654–74.

Hutton, Jane, ed. *Landscript 5: Material Culture: Assembling and Disassembling Landscapes.* Berlin: Jovis, 2018.

Jackson, Peter. "Commercial Cultures: Transcending the Cultural and the Economic." *Progress in Human Geography* 26, no. 1 (2002): 3–18.

Jackson, Peter. "Rematerializing Social and Cultural Geography." *Social & Cultural Geography* 1, no. 1 (September 1, 2000): 9–14.

Kennen, Kate, and Niall Kirkwood. *Phyto: Principles and Resources for Site Remediation and Landscape Design*. New York: Routledge, 2015.

Kimmerer, Robin Wall. *Braiding Sweetgrass: Indigenous Wisdom, Scientific Knowledge and the Teachings of Plants*. Minneapolis: Milkweed Editions, 2015.

Krausmann, Fridolin, Simone Gingrich, Nina Eisenmenger, Karl-Heinz Erb, Helmut Haberl, and Marina Fischer-Kowalski. "Growth in Global Materials Use, GDP and Population During the 20th Century." *Ecological Economics* 68, no. 10 (August 15, 2009): 2696–705.

Margolis, Liat, and Alexander Robinson. *Living Systems: Innovative Materials and Technologies for Landscape Architecture*. Basel and Boston: Birkhäuser, 2007.

Marjaba, Ghassan and Samir Chidiac. "Sustainability and Resiliency Metrics for Buildings: A Critical Review." *Building and Environment* 101 (May 2016): 116–25.

Martinez-Alier, Joan. *The Environmentalism of the Poor: A Study of Ecological Conflicts and Valuation*. Cheltenham: Edward Elgar, 2002.

Marx, Karl, and Ernest Mandel. *Capital: A Critique of Political Economy*. Translated by Ben Fowkes. New York: Penguin, 2004.

McNeill, J.R. *Something New Under the Sun: An Environmental History of the Twentieth Century World*. New York: WW Norton, 2001.

Meyer, Elizabeth. "Site Citations: The Grounds of Modern Landscape Architecture." In *Site Matters: Design Concepts, Histories, and Strategies*, edited by Carol J. Burns and Andrea Kahn, 93–130. New York: Routledge, 2005.

Mitchell, Don D.M. "Cultural Landscapes: The Dialectical Landscape – Recent Landscape Research in Human Geography." *Progress in Human Geography* 26, no. 3 (June 1, 2002): 381–9.

Moore, Jason W. *Capitalism in the Web of Life: Ecology and the Accumulation of Capital*. New York: Verso, 2015.

Mukerji, Chandra. "Space and Political Pedagogy at the Gardens of Versailles." *Public Culture*, 24, no. 3 (2012): 509–34.

Muradian, R., and S. Giljum. "Physical Trade Flows of Pollution-Intensive Products: Historical Trends in Europe and the World." In *Rethinking Environmental History: World-System History and Global Environmental Change*, edited by A. Hornborg, J.R. McNeill, and J. Martinez-Alier, 307–26. Walnut Creek: AltaMira Press, 2007.

Palmer, Laurie. *In the Aura of a Hole: Exploring Sites of Material Extraction*. London: Black Dog Publishing, 2015.

Pollan, Michael. *The Omnivore's Dilemma: A Natural History of Four Meals*. New York: Penguin, 2007.

Sanderson, Eric W. *Mannahatta: A Natural History of New York City*. New York: Harry N. Abrams, 2013.

Smithson, Robert. "Frederick Law Olmsted and the Dialectical Landscape." In *Robert Smithson: The Collected Writings*, edited by Jack Flam, 157–71. Berkeley: University of California Press, 1996.

Smithson, Robert. "A Provisional Theory of Non-Sites (1968)." In *Robert Smithson: The Collected Writings*, edited by Jack Flam, 364. Berkeley: University of California Press, 1996.

Smithson, Robert. "A Sedimentation of the Mind: Earth Projects." *Art Forum*, September 1968, republished in *Robert Smithson: The Collected Writings*, edited by Jack Flam, 100–13. Berkeley: University of California Press, 1996.

Steinberg, Ted. *Gotham Unbound: The Ecological History of Greater New York*. New York: Simon & Schuster, 2015.

Sundberg, Juanita. "Decolonizing Posthumanist Geographies." *Cultural Geographies* 21, no. 1 (2004): 33–47.

Sustainable Sites Initiative. "Guidelines and Performance Benchmarks, 2009." Pp. 124–38. Accessed October 12, 2018, https://digital.library.unt.edu/ark:/67531/metadc31157/m2/1/high_res_d/Guidelines%20and%20Performance%20Benchmarks_2009.pdf.

Thompson, William, and Kim Sorvig. *Sustainable Landscape Construction: A Guide to Green Building Outdoors*. Washington, DC: Island Press, 2000.

Treib, Marc. "Must Landscapes Mean?: Approaches to Significance in Recent Landscape Architecture." *Landscape Journal* 14 (January 1, 1995): 46–62.

Whatmore, Sarah. "Materialist Returns: Practicing Cultural Geography in and for a More-Than-Human World." *Cultural Geographies*, 13 (2006): 600–9.

Who Builds Your Architecture?: A Critical Field Guide. 2017. "WBYA_Guidebook_spreads.Pdf." Accessed October 21, 2018, http://whobuilds.org/wp-content/uploads/2017/02/WBYA_Guidebook_spreads.pdf.

Wilkinson, Bruce H. "Humans as Geologic Agents: A Deep-Time Perspective." *Geology* 33, no. 3 (2005): 161–4.

Figure 1.1(A) Detail of guano deposit and settlement, Chincha Islands, Peru, 1862

Source: Courtesy of New Bedford Whaling Museum.

Figure 1.1(B) Detail of Sheep Meadow looking southwest, Central Park, circa 1905

Source: Photograph by William Hale Kirk (© William Hale Kirk/Museum of the City of New York).

Chapter 1

Inexhaustible Terrain
Guano from the Chincha Islands, Peru, to Central Park, 1862

To a blade of grass, access to nutrients means everything: nitrogen accelerates growth and makes for bright green color, phosphorus builds stronger roots and winter toughness, and potassium stiffens blades to make a sturdier surface. To Frederick Law Olmsted, carefully fertilizing and improving the soils of Central Park was essential to create the sweeping pastoral landscape that it would become famous for. The success of the turf would impact the beauty and success of the park as a whole.[1] "The surface . . . can hardly be made too fine or too smooth and even," Olmsted wrote of the grasses in Central Park, "nor can the turf afterwards be kept too free of any plant except the grasses, nor can the grass be kept too short, or be too smoothly rolled."[2] Olmsted and Calvert Vaux conceived of Central Park as a pastoral retreat for frazzled New Yorkers alienated from the countryside.[3] The park looked like a farm, but it also operated like one. Olmsted experimented with new fertilizers being used in farms at the periphery of the expanding metropolis, including horse manure from street sweepings, "night soil" from Manhattan's privies, manufactured *poudrette*, and guano. Of the fertilizers applied to Central Park at the time, guano (desiccated seabird excrement) was the most potent, exotic, and novel. Unlike composted manures and fertilizers incorporating local industrial byproducts, guano was imported from Peru and the South Pacific.

With thousands of miles between them, the Chincha Islands of coastal Peru and Central Park are improbably connected. This chapter follows a trace amount of Peruvian guano applied to the initial soils of Central Park in the early 1860s. While small in volume, this particular fertilizer application reflects guano's transformative role in farms surrounding New York City and other industrializing metropolises in North America and Europe. As farmers shifted from self-sustaining to industrial models, and substituted more processed and mined fertilizers for the composted animal manures traditionally used, nutrient cycles expanded from the scale of the farm to the urban region to the planet. In the quest to recharge depleted soils in one place, ecological deposits and humans were exploited and exhausted in another. The case of guano in Central Park more broadly reflects the growing metabolic rift of the nineteenth century, as expanding nutrient cycles and enslaved human labor were invisibly, yet directly, linked to the public landscape.

"Nature's Reciprocity System"

Before long, Olmsted and Vaux argued, Manhattan would be crowded with buildings and noisy, noxious streets.[4] In their 1857 winning competition entry for the design of Central Park, Olmsted and Vaux stated that a great purpose of the park was to "supply to the hundreds of thousands of tired workers, who have no opportunity to spend their summers in the country, a specimen of God's handiwork."[5] The park had to counteract the damaging forces of rapid urban growth. It wasn't enough to just introduce plants in the park site; trees in a row or pots of roses could be found in any urban garden. It wasn't enough to just improve poor air quality, although that was definitely important. By offering natural "scenery," the design could impact people's minds and spark their imaginations.[6]

Pastoral scenery could do this best. Olmsted pictured a meadow where shadows on the far ground would produce illusions of variation and distance, and where trees and topography would hide the buildings along the park's boundaries. Pastoral scenery referred directly to working agricultural landscapes – and the type that many of New York City's new residents had migrated from. No longer in fields, citizens now worked in factories estranged from the land and its cycles. Visitors experienced the remediating effects of the different environments, but they would also see the park working as a farm. At Central Park's Dairy, children could get affordable, fresh milk of higher quality than that more commonly available from brewery cows.[7] The public could watch hay and rye crops growing and watch as workers periodically scythed it down to feed the park's animals. The Sheep Meadow – the gem of the design's pastoral landscape – was the park's most literal agricultural simulacra.

Figure 1.2 **Sheep Meadow looking southwest, Central Park, circa 1905**

Source: Photograph by William Hale Kirk (© William Hale Kirk/Museum of the City of New York).

Figure 1.3 **Robert L. Bracklow, Sheep in Sheep Meadow Central Park, circa 1890–1910, glass negative, 5 in. × 7 in.**

Source: Robert L. Bracklow Photograph Collection, 1882–1918 (bulk 1896–1905), nyhs_pr-008_66000_1000, Photography © New York Historical Society.

The Sheep Meadow, originally named the Parade Grounds to satisfy competition requirements for military programming (and intermediately called "The Green"), looked like a real pasture. If the grasses made a pastoral scene, a flock of Southdown sheep activated the agricultural stage from 1864.[8] Roving collies, protecting the flock from stray dogs, offered a living "object lesson" for kids who otherwise had no exposure to farm life.[9] Visitors could watch a Parks Department employee lead the flock to the Sheepfold every evening; they could purchase wool at annual auctions held by the Parks Department.[10] Sheep were valued as rural curiosities, but also as trimmers and fertilizers. Sheep graze close to the plant roots, "closer than any lawn mower yet devised," and produce a short-cropped, tough turf, as one *New York Times* author noted.[11] Their droppings helped the grass grow quickly. While the sheep produced a symbolic image of a farm, their role in the ecological functioning of the soil and grass was part of this image.

As the new contours of Central Park were formed, workers mixed composted horse manure and night soil into the top layers of the soil, distributing it at 300 or 225 cartloads per acre, respectively.[12] Horse and human manures were abundant and cheap, and so they were the primary soil amendments. Manures give plants access to nutrients slowly by changing the quality of the soil; they aggregate soil particles, which makes them spongier and absorptive and provides conditions for microorganisms to thrive. With the help of specialist fungi, bacteria, and other organisms, manures are transformed into the nutrients that plants require.

Guano, in contrast, gives plants more highly concentrated nutrients, immediately, without transforming soil quality or its microbiota. Much more potent than farm animal or human manures, guano was applied at five bushels to the acre, or approximately half a cartload per acre.[13] Guano, like other highly concentrated fertilizers, was used primarily to accelerate the growth of

Figure 1.4 **Construction activity in Central Park, 1858**

Source: Lithograph by George Hayward (Picture Collection, The New York Public Library, Astor, Lenox and Tilden Foundations).

annual cover crops, which were planted and then turned back into the soil.[14] Perennial grasses such as Red-top grass (*Agrostis vulgaris*) and English Lawn grass (*Agrostis stolonifera*) were then sown with white clover in the spring. In the first year of growth, tender perennial grasses needed to be protected, and so annual crops were inter-seeded. In the summer, millet seed was added, as its quick growing stalks could shade the young grasses from the sun. In the fall, rye was sown in to help shield the grasses over the winter.[15]

While the Central Park Annual Reports provide few details on practices as specific as fertilizer application, the Seventh Report goes into detail about a range of fertilizer types used to prepare the soil in different areas of the park. The report describes how guano was applied in the 11.5 acres east of the Old Reservoir (the present-day Great Lawn) and the 2 acres west of it, at 500 pounds per acre. Guano in this case was a mix of one part Peruvian to three parts Baker Island guano ("American" guano from the South Pacific), mixed with eight pounds of salt per 100 pounds of guano.[16] Few other areas receive specific mention about their fertilization, suggesting that these parcels were of particular interest. Across the park, once perennial lawns were established, Park workers cut the grass twice weekly and monitored the changing quality of the turf as it confronted the waves of visitors. "The perfection of such meadow and glade surfaces is found in nature only in the spring, when the turf is still short and growing evenly," Olmsted wrote, "but by shaving the grass at frequent intervals, this perfection can be nearly maintained through the summer."[17] Biologically speaking, getting this "perfection" involves aborting the plant's sexual reproduction, preventing the setting of seed and instead stimulating the plant to reproduce rhizomatically, rendering a thicker, denser mat.

Shortly after opening to visitors, a regulation prohibited the public grazing of animals (apart from the park's own flock) on the park's grounds.[18] Protecting

Figure 1.5 Elizabeth Ransom, 'Keep off the grass' sign in
Central Park near the Reservoir, 1891, Glass plate negative

Source: Elizabeth Ransom Photograph Collection, 1891–1892,
nyhs_PR237_s-01_b-01_f-03_62684, Photography © New York
Historical Society.

the turf was critical, yet this paradoxically excluded people from the very thing
that was meant to entice them. Frail, newly seeded ground provided an excuse
to temporarily limit access; however, the prohibitions eventually became per-
manent. Throughout the week visitors were prohibited from walking in the
Sheep Meadow except in certain areas. Visitors were allowed to picnic in the
meadow on Saturdays, although this excluded the poorest New Yorkers, who
were typically only free from work on Sundays.[19] Not only would public use
of the turf effectively destroy it, public activities threatened the quiet, reflective
qualities that Olmsted and Vaux desired for the meadows. Central Park's green
meadow scenery, supposedly constructed for the masses, was off limits. This
contradiction between grass and its users foreshadows the 100-year development of a now multi-billion-dollar turf industry, which I will return to later on.

Although visitors could witness farm activities in Central Park, most experienced it as a site of leisure rather than labor, a space of respite in a dense
urban environment. While the park sold milk, hay, and wool, among other
products, its major economic value came from its influence on the real estate
market. Central Park's pastoral aesthetic produced value for the exploding real
estate speculation developing along its borders. By formalizing a large segment
of the city that had been informally settled and used by a range of inhabitants,
Central Park ensured the success of higher-end residential development.[20] By
occupying such a large tract of land, it increased the value of surrounding
landholdings. Capitalists in New York City benefited from a new version of
the agricultural landscape: where wealth had once been generated through the
ownership (informal or otherwise) of fields and their fruits, it could now be
gained through proximity to a symbol of agricultural retreat.

Central Park was not only a potent symbol of the agricultural landscape,
it also functioned like one. Like farms in the region, the park processed a spectrum of urban wastes through its soil, both as fertilizer and waste-management

strategy. The types of fertilizers available at the time of the park's construction reflect the re-scaling of urban–rural relationships that was well underway. From manure to manufactured *poudrette* to highly concentrated imported guano, these materials – as well as their transformation and transportation – demonstrate the shift of agriculture towards an industrial model.

New York City's population had exploded in the mid-nineteenth century as rural migrants and European immigrants arrived en masse, shaping the city both as a powerful new labor market and consumer base. With the unprecedented construction of new buildings, infrastructure, and industry came unforeseen quantities of waste from humans, horses, and factories. These wastes constituted an exciting, bountiful stock of cheap nitrogen and phosphorus, and were immediately recognized as an asset for farmers growing food on the periphery. At the same time, overflowing privies, mountains of horse manure, and industrial effluents presented a new municipal dilemma of unforeseen proportions.

Even the land slated for Central Park contained urban wastes that could be used to fertilize. Uptown, beyond the close watch of the city's regulators, communities thrived with access to space – and more importantly, soil. Far from an uninhabited landscape, the future park site was home to a heterogeneous mix of settlements, small industries, market gardens, and institutions. Communities took advantage of the semi-rural land for cultivation and animal husbandry, supplementing their income and putting food on the table with vegetables, meat, and products from the land. Like any agricultural landscape, organic wastes produced on site would have been cycled back through the soil as a matter of practice. In Seneca Village, just north of the future Sheep Meadow, residents of the first community of African American landholders in the United States tended gardens and kept animals in barns. At the southeastern corner of the park, a settlement of Irish immigrants raised pigs, while German immigrants developed extensive produce gardens in the rocky soil.[21] Small industries recycled waste products and would have released nutrients to the land; for example, bone-boiling plants processed waste bones to make soap and other products, all the while leaking byproducts into the site. With the eviction of these residents went the last remnants of agricultural Manhattan, yet their activities were recorded in the soil. Inspecting the park site before construction, Olmsted noted how the evicted offal shops and market gardens had greatly improved soil fertility. He specified that this valuable earth be relocated to areas with nutritionally deficient soil.[22]

Of the materials first introduced into the park's soil to fertilize, composted horse manure and night soil were local, abundant, and could sometimes be had for free. As early as 1818, horse manure was recognized as an infrastructural waste problem in New York City.[23] Horses thronged streets, piles of manure engulfed valuable land, and "dirt carters," those who transported manure to larger composting lots, were regulated by the city. Airborne particles and streams of liquid manure permeated the urban environment.[24] Manure could be offloaded to Department of Parks composting lots, but only if it was "delivered free" since there were plenty of stables wanting to deposit their manure

Figure 1.6 **Horse manure in Manhattan, 1893**

Source: George E. Waring, *Street-cleaning and the Disposal of a City's Wastes*. New York: Doubleday & McClure Publishers, 1897, pp. 8–9.

somewhere.[25] The same horses that dropped the offending matter powered the elite's carriages, which both served as an important display of wealth for their owners and also physically separated the classes throughout the park.[26] In 1869, thronged carriage drives in the park produced more than 3,000 cartloads of manure – a volume so great that no additional fertilizers had to be purchased for park use.[27] Not only was this composted manure applied within Central Park, but it served as a source for top-dressing the city's many other parks.[28]

Similarly, residential privies brimming with "night soil," or human feces were a significant waste-management problem. Growing aware of the links between poor sanitation and disease, the city encouraged independent "night-scavengers" to gather night soil and sell it to agricultural regions in need of large quantities of fertilizer. This privatized system of waste removal delivered an essential service, while providing jobs to many low-income residents.[29] In 1866 alone, some 60,000 cartloads were transported from New York City homes to river docks and the periphery, some notoriously dumped right into the river or applied directly to regional agricultural fields.[30] Night scavengers also delivered carts of night soil to Central Park in its first years, where it was dumped into broad pits, mixed seven parts earth to one part night soil, and left to ferment for a few weeks before application.[31] New York City's sewage system was developing haphazardly, and so night soil was plentiful until a decade after Central Park was constructed.[32]

Agriculture societies and municipal committees in New York City debated the implications of the rapidly growing waste trade and the new market of high-potency fertilizer products.[33] How could new urban waste streams support intensifying agricultural production? What biological and economic stakes were involved with introducing imported fertilizers? Two figures involved in the development of Central Park spoke out. Egbert Viele, who had designed the original drainage scheme for the park, chaired the city's newly formed "Committee on Civic Cleanliness," launched in 1859. In a report detailing recommendations for sanitary reform, Viele emphasized the cycling of urban wastes as an urgent matter of economy and health. While new sanitary reform called for the rapid removal of offal, night soil, and manures from city streets, Viele argued that these were valuable and shouldn't simply be disposed of. Such material "properly belongs to agriculture" and so should be deposited outside of the city limits: "in time a most valuable accumulation of the richest kind of fertilizing material would yield a revenue to the city."[34]

George E. Waring, the park's agricultural engineer, advocated for the use of night soil in Central Park and also promoted the Earth Closet as a means to collect and recycle night soil as high-quality compost. "Our bodies have come out of our fertile fields," Waring wrote, "our prosperity is based on the production and the exchange of the earth's fruits; and all our industry has its foundation in arts and interests connected with, or dependent on, a successful agriculture."[35] He mourned the loss of human waste to the sea through sewers, estimating that $50 million worth of excrement (or phosphoric acid, based on bone flour prices) was lost annually in the United States. But more urgent than lost money was the existential dilemma of losing the capacity to produce. "If mill streams were failing year by year, and streams were yearly losing force, and the ability of men to labor were yearly growing less," Waring cautioned, "the doom of our prosperity would not be more plainly written than if the slow but certain impoverishment of our soil were sure to continue."[36]

To address both mounting urban waste and also enrich agricultural soil with local fertilizers, local manures seemed to make the most sense. These wastes – blocking traffic, overflowing stables and privies – had originated as crops harvested from peripheral agricultural lands and were processed through the intestines of the metropolis in a fairly localized ecological loop. Historian Richard A. Wines describes how the region "in essence became a huge recycling system,"[37] or as one contributor wrote in *The Horticulturalist* in 1846, "nature's reciprocity system."[38] Raw manures, however, were heavy and burdensome and so only cost-effective to use within a certain distance from their source.[39]

As farms and their markets grew further apart, practices known to maintain soil fertility, such as crop rotation and application of local manures, were abandoned in favor of mechanical farming equipment and the use of more stable, transportable, and homogeneous fertilizers. Farmers became accustomed to purchasing fertilizers, and crossed what Wines calls the "psychological boundary between self-sufficient and capitalistic farming."[40] The move from bulky manures towards fertilizer "products" further shifted nutrient cycling towards a commodity market. Poudrette, dried night soil mixed with plaster based on a French recipe, for example, was manufactured in fertilizer plants and promised a lighter weight, more potent, dry product that could be more easily transported, stored, and applied.[41] Other urban wastes, such as recycled bones from the sugar industry, blood from slaughterhouses, and ammonium sulfate from gas production, were useful precursors to more processed fertilizer products, yet were difficult to obtain consistently in large quantities. The

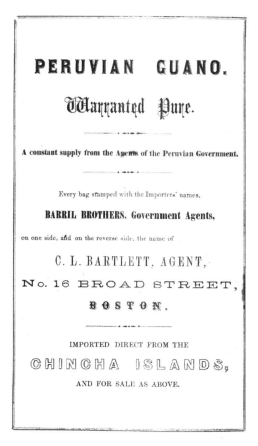

Figure 1.7 **Advertisement for guano from the Chincha Islands, 1860**

Source: Charles L. Bartlett, *Guano: A Treatise on the History, Economy as a Manure, and Modes of Applying Peruvian Guano.* Boston: C.L. Bartlett, 1860.

burgeoning fertilizer industry turned away from recycled sources and towards mineral phosphates.[42] Inputs shifted from local, recycled, organic wastes to non-renewable and distant mineral sources – from recycled to mined. And the world turned towards Peruvian guano.

Guano was a gateway for farmers moving towards industrialized agriculture. Adopted earlier by British farmers, guano found a commercial market in the New York region by the 1840s, and widespread use took off in the 1860s. Guano was an easily transportable, quick acting, highly concentrated source of nitrogen. It was a mineral-like product, yet technically a manure; it also smelled like ammonia, and was therefore familiar to farmers accustomed to spreading animal waste on farms.[43] It outperformed other commercially available fertilizers, and stood out for its consistent performance within a very inconsistent market. "Among animal manures it clearly claims the first place," Charles L. Bartlett's 1860 text quoted agricultural chemist Dr. Von Martius in his treatise on guano, "It is five times better than night soil and also very superior to the French poudrette."[44] The title page to Bartlett's treatise features an epigraph: "Guano, though no saint, works many miracles." And while opinions varied about all of the new fertilizers, guano was consistently praised: "No scientific man, indeed, has ever expressed an opinion against the extraordinary and unequaled fertilizing properties of Peruvian guano," Bartlett wrote.[45] For those wary of introducing a foreign material and disconnecting from a local organic waste cycle, it was possible to view guano as simply an extension of that system.[46] Urban wastes washed into the oceans, and ocean nutrients fed sea life, which in turn fed guano birds. It could be conceived of as a complete circuit of global nutrient cycling, and justified as necessary to replenish local soils given the growing exportation of crops.[47]

Others viewed guano as a paradigm shift with treacherous consequences. German chemist Justus von Liebig, who demonstrated the significance of soil nutrients on plant growth, opposed the growing dependence upon chemical fertilizers. Liebig warned,

As his forefathers believed that the soil of their fields was inexhaustible, so the farmer of the present day believes that the introduction of manures from abroad will have no end. It is much simpler, he thinks, to buy guano and bones than to collect their elements from the sewers of towns, and if a lack of the former should ever arise, it will then be time enough to think of a resort to the latter. But of all the erroneous opinions of the farmers this is the most dangerous and fatal.[48]

Liebig's work on soil fertility was widely read, influencing, ironically, the future development of synthetic fertilizers, but also Karl Marx's theory of social metabolism. In *Capital*, Marx argued that urbanization under capitalist production disturbed fundamental relationships between humans and the earth. Displacing nutrients in the form of food and clothing from the countryside to the city, he wrote, "hinders the operation of the eternal natural condition for the lasting fertility of the soil. . . . All progress in capitalist agriculture is a progress in the art, not only of robbing the worker, but of robbing the soil."[49] For Marx, the disruption of local nutrient cycling was a "rift" of massive consequence, and the source of the inevitable ecological crises that capitalism produces, the "metabolic rift."[50] Soil robbery happens when nutrients are not cycled back into agricultural land after being consumed in adjacent cities, but also at the scale of the globe, through colonization.

Industrializing cities were powered by produce and nutrients stolen from colonial and imperially controlled land, never to be returned. This "ecological imperialism" produces ecologically unequal exchange, where the exploitation of natural resources and labor in the "periphery" to supply the "core," or industrialized/ing centers, and peripheral countries are disproportionately environmentally degraded, undercompensated, and robbed of their capacity (in terms of natural resources and labor) to thrive. For Brett Clark and John Bellamy Foster, guano is a prime case of the emerging global metabolic rift, as

Britain and the United States (among other countries) drew the fertilizer from Peru to improve soil exhausted by unsustainable agricultural practices.[51] And in the United States, the depleted farms of the Southeast were ravenous.

Exhaustion

When Olmsted specified guano for Central Park, he did so from experience and out of professional curiosity. Just a few years earlier, his book *Journey in the Seaboard Slave States: With Remarks on Their Economy* was published.[52] Commissioned by the *New York Daily Times* to report on the workings of American slavery, Olmsted wrote a series of personal account articles (later compiled as a book) detailing his visits to farms in Washington, DC, Maryland, Virginia, the Carolinas, Georgia, Alabama, and Louisiana. As a budding farmer, Olmsted's journalistic lens was focused on the conditions of slavery as well as agricultural techniques. While Olmsted would later be associated with an abolitionist perspective, his report reflects a detached tone, revealing the atrocities of slavery in sober language alongside inquisitive, detailed descriptions of farming practices.

Generations of intensive tobacco and cash crop farming had left soils throughout the American South exhausted. Rather than rotate crops, farmers tended to "rest" spent soils and instead cultivate more and more land, which lead to widespread soil erosion, gullying, and impoverishment.[53] In the years leading up to the Civil War, poor agricultural production threatened the South's economic capacity for secession and the right to maintain an economy based largely on the violent exploitation of slavery. When guano arrived in

Figure 1.8 **Illustration from Frederick Law Olmsted,** *A Journey in the Seaboard Slave States*, **New York: Dix & Edwards, 1856, p. 387**

Baltimore's ports, it offered desperate farmers hope for their depleted soils. One South Carolinian farmer reflected: "whether it [guano] will work out the problem we have been so ardently bent upon or only hasten the exhaustion of an already impoverished region is yet to be determined."[54]

Olmsted was impressed with the widespread and successful use of guano, and he documented different applications throughout his travels. In Washington, DC, he noted that guano had "immensely improved" one farmer's wheat crop, and discussed his own successes with the fertilizer.[55] While he was excited about guano's power to raise crop yields, Olmsted also cautioned against its careless use: "Where ignorantly or improvidently employed, with a thought only of immediate returns, it will probably lead to a still greater exhaustion of the soil, and lessen the real wealth of the poor farmer."[56] Guano, in Olmsted's opinion, was an amendment of great value, but "real wealth" came from a longer-term consideration of soil health.

Farmers told Olmsted their belief that, in addition to their problems with exhausted soil, their enslaved workers were lethargic. "Mr. W." claimed that slaves worked at half their ability: "They could not be made to work hard: they never would lay out their strength freely." Their will to hold some strength back, Olmsted's informants claimed, led to depleted soils, poor crop yields, and economic instability. As Jennifer C. James writes, the characterization of workers as holding energy back complicates typical stereotypes about laziness; instead she writes, the slaveholder "bestows the slaves with magically self-renewing powers which allow him to extract their labor without limit or guilt. They can never be overworked, nor can they ever work enough."[57] Not only was the enslaved "indolence" a frustration for owners, it worked against the "civilizing" values of cultivation. "If you could move all the white people from the whole seaboard district of Virginia and give it up to the negroes that are on it now, just leave them to themselves," one of Olmsted's sources predicted, "in ten years time there would be not an acre of land cultivated, and nothing would be produced, except what grew spontaneously."[58] From that position, slavery was a virtuous force preventing the further obliteration of the soil. Olmsted likens this fear that "Africanization" of the land would lead to a desolate, wild, and barren landscape, with Virginia politician Willoughby Newton's belief that without guano, the land would be fast deserted: "I look upon the introduction of guano . . . in the light of a special interposition of Divine Providence, to save the northern neck of Virginia from reverting entirely to its former state of wilderness and utter desolation."[59] According to this logic, like slavery, guano could civilize the land; without it, it would return to wilderness.

In the eyes of slaveholders, guano and slaves both added value to the fields; however, guano's "strength" had seen no match, and slavery might soon be abolished. The editor of the *Southern Cultivator* wrote in 1853: "farmers should use more guano . . . instead of investing their money in more Negroes."[60] Farmers' acclaim for guano, Jennifer C. James argues, was also a way to undermine the slaves' work; while they praised guano for its "work," they negated the enslaved workers' role in growing and processing the bounty that the fertilizer would produce.[61] After the Civil War, realizing the costs of paying for labor, farmers began to lean on guano where human labor couldn't be afforded. Willoughby Newton acknowledged this change after emancipation: "No man can suppose it possible to cultivate profitably, poor land with hired labour; and under our new system fertilizers must be used to a much greater extent than formerly."[62]

As abolition approached in the United States, new systems of exploitation began to fill the dawning labor gap. American abolition threatened the supply of cheap cotton that had been flowing from the American South, and so capitalists went searching for other locations around the world, including Brazil and Peru, where they could establish lucrative plantations, often with forced labor.[63] In the years leading up to the abolition of American slavery, the media took note of these global labor dynamics. The abolitionist *Frederick Douglass' Paper* reported in 1855 that "a new slave trade is going up in the world."[64] "The revelation is frightful"; the article detailed the cruel treatment of Chinese workers in the Peruvian Chincha Islands, under British and Peruvian control.

Excretion

Approaching from the coast of Pisco, Peru, the Chincha Islands appear like three giant barnacles latched onto the horizon. The islands, each around one mile across, are part of a guano-coated archipelago that runs along the country's coast. Jhuneor Paitan Ñahui, a specialist in the Guano Islands and Capes National Reserve (*Reserva Nacional Sistema de Islas Islotes y Punta Guaneras*) at SERNANP, Peru's protected areas environmental agency, has agreed to take me to visit the islands, and we speed westward by boat to reach them.[65] Before my trip to the Chincha Islands, I had seen a striking lithograph from 1865 showing a nearly similar vantage point with a very different view – towering islands, battered in profile, surrounded by a gang of ships jockeying for a position underneath the docks. These cliffs of "white gold" dwarfed the massive ships below, offering a seemingly endless supply of the highest quality fertilizer in the world.

Just a decade before that lithograph was made, New York–based naturalist and voyager George Washington Peck, who wrote extensively of his experiences in the Chincha Islands, described the same scene in words:

> Imagine the Andes and the Pacific in one view – the islands with their precipitous walls indented with immense caves, and surrounded by fantastic rocks, fringed with foam – the pure ocean air – the myriads of sea birds – the shipping – the schools of sea lions – and almost always, far or near upon the blue waste – the spout of whales, and the white sails of ships coming or departing – altogether the scene is full of exhilaration and excitement.[66]

At the height of the Peruvian guano trade from the beginning of the 1840s through to the end of the 1870s, millions of seabirds, thousands of enslaved

Figure 1.9 **Plan and perspective of the Chincha Islands, 1865**

Source: Lithograph by Mariano Felipe Paz Soldan (Atlas map, 35 cm × 19 cm. David Rumsey Map Collection).

Chinese people, and scores of ships bound for New York City, Baltimore, and European ports swarmed the three islands. The "mountains" were composed of seabird guano: deposited over millennia, yet depleted within decades. The ships are gone now, and the profile of the islands is unrecognizably shallow.

Today, just one man lives on each island – alone with a lot of birds. Jorge Tarazona Paredes lives on Isla Chincha Norte, the northernmost, where he monitors the bird populations, hosts scientific researchers, guards against unauthorized fishing and poaching, and waits out the years before the next round of guano harvesting.

Jorge lowers a slippery rope ladder into our boat so that we can scramble up to the dock suspended over churning waters. Isla Chincha Norte hasn't had any human guests for the past month, and Jorge seems to welcome our company. Tethered to the island bedrock with cables, the main dock leads us through a cluster of small buildings and structures. Once-active docks today look like grounded pirate ships, their masts covered in birds. Today they are the primary visual evidence of humans in a landscape dominated by other life. To our left, a long wooden building with a broad porch reads "ISLA CHINCHA NORTE" in light blue letters, large enough to see from the water. It had been the busy headquarters of the guano administration, but now, more like a quiet roadside motel, it is Jorge's living quarters and office. At the rear of the building is a small sink and trellis supporting a small vine – the only plant that I see on the islands. To our right, a small outbuilding's windows have been boarded up and covered with woven guano bags, each pictured with a guanay cormorant perched on a guano bag. Meanwhile, living guanays cover the building as dense as roof tiles. Beyond, an empty soccer pitch is surrounded with a low brick enclosure to keep balls from rolling out into the arid expanse.

Figure 1.10 **Docks, North Chincha Island, 2014**

Figure 1.11 Guanays and guano bag window coverings, North Chincha Island, 2014

Figure 1.12 Main administrative building, entrance bridge, and soccer pitch beyond, North Chincha Island, 2014

Figure 1.13 **Former hospital with Jorge Tarazona Paredes and Jhuneor Paitan Ñahui in foreground, North Chincha Island, 2014**

A crater field of guanay nests blankets the landscape as far as I can see, and makes me feel that we are walking on the moon. The same round nests inspired Indigenous people from the Pisco region to use the word *quillairaca* or "the moon's vagina" when referring to the islands.[67] Periodically, birds storm overhead and resettle on different surfaces in the distance, but the areas we walk through are quiet for the moment. Three species – the guanay cormorant (*Phalacrocorax bougainvilli*), Peruvian booby (*Sula variegata*), and Peruvian brown pelican (*Pelecanus occidentalis thagus*) – are responsible for the majority of the guano mass, but many others move between islands throughout the day, fishing, preening, nesting. Speckled black hillsides reveal themselves to instead be white as flocks of guanays take off. From the open planes exposed to the sun, to the undercut caves at the water's edge, animals clamor.

Peck described the life teeming in the many surrounding caves:

> Far within, the dark dripping ledges may be seen covered with nests and birds wherever nests can be stuck or birds stand, and along with the wind and spray that rushes out as the waves advance, come the hoarse cries of penguins and sometimes the roar of sea lions.[68]

On the land, penguins mingle with sea lions, pelicans, and terns, apparently unbothered by each other and focused instead on the water. Cacophonous

Figure 1.14 Guanay nest landscape, North Chincha Island, 2014

Figure 1.15 Guanay nest detail, North Chincha Island, 2014

Figure 1.16 **Guanays, North Chincha Island, 2014**

terrestrial life signals an intensely productive and diverse sea-life below. Positioned within the powerful Humboldt Current – which carries cold water up the coast of South America – the islands are immersed in an underwater tumult of debris. Close to central Peru, where the Chinchas lie, a powerful upwelling in the current draws great amounts of organic matter up from the ocean floor. Phytoplankton and other microorganisms thrive in that environment and are a banquet for small fish.[69] Anchovies (*Engraulis ringens*) are the primary food for guano seabirds, and their particularly nutrient-rich bodies produce guano's high nitrogen and phosphorus content.

The island is a super-active volume. Granite bedrock forms a stable plateau above the sea, but the mass that lies on top of it continually builds. Birds and sea lions that bask on its beaches, nest on its peaks and inside caves, and dive for fish at its margins literally become the island. Carcasses, eggs, bones, and sea lion teeth, cemented in layer upon layer of bird feces, change from a conglomerate of biotic materials into homogenous sedimentary matter. Over time, the weight of accumulating guano compacts and concentrates the material to a hard-packed mass, one that Peck described as having the consistency of Castile soap.[70] Thick, white, excrement itself, when first laid down, is already a "product"; anchovy bodies consumed, transported, and excreted on the land, powered by the digestive tracts and wing-stroke of sea birds, are folded into the island to be further compacted over time. A multi-colored stratigraphy develops over time; young guano, deposited on top of the pile, is practically white, but as it becomes buried by more material, it shifts to gray and then to reddish brown.[71]

Like a glacial mass upon bedrock, the guano cap forms in response to, but independently from, its substrate. Along with the process of continual deposition and build-up of excrement, other animal and weather agents constantly reshape the guano cap. Wind carves the mass over time, hewing an aerodynamic profile. For most other guano islands, rain would erode the deposits, but as it happens, there is little to no rain on the Chincha Islands – the secret to their domination of the industry. Bird species burrow in it and sea lions slide across it, continuously shaping the guano mass contours.[72] And, of course, the

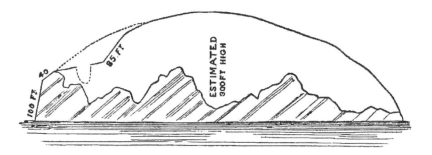

Figure 1.17 **Guano section, North Chincha Island**

Source: Solon Robinson, *Guano: A Treatise of Practical Information for Farmers*. New York : Solon Robinson, 1853, p. 78.

cap is shaped by human exploitation. While small-scale harvesting occurred initially just at the accessible margins, large-scale operations later completely erased it.

It is difficult to estimate how the guano cap grew, given that it is such a dynamic landform and has had few human chroniclers. G. Evelyn Hutchinson, the influential modern ecologist and author of a definitive text on guano, *The Biogeochemistry of Vertebrate Excretion*, studied various accounts of guano strata to estimate rates of guano accumulation on the Chincha Islands. He determined that 1.9 centimeters (or three-quarters of an inch) a year was a reasonable deposition estimate, dating guano caps of the nineteenth-century guano boom to the fifth century.[73] From where we stand facing the crest of the island, guano covers most surfaces, but it is relatively shallow on the bedrock. I try to imagine the guano peaks that once buried this island. I identify a spot 150 feet above my head, the estimated height of the guano hill that Peck perched upon in 1854. From the ground to the point in the sky, I visualize layer upon layer of annual depositions covering the entire island. The mental image is sublime, a volume of truly awesome biological processes. Even if I could grasp the scale of the guano mountain, the numbers of birds and fish that made and became it are baffling. Where there is now only white sky and the circling of birds once stood a monument to accumulation, the trophy of an exceptionally vital ecosystem.

The mountain turned from an ancient accretion into a lucrative material stockpile. Over four decades following 1840 – the so-called "guano age" – some 10 million tons of guano, valued at half a billion dollars, were exported from the Chincha Islands.[74] A. J. Duffield described the visible effects of this extraction on the islands in 1877:

When I first saw them twenty years ago, they were bold, brown heads, tall, and erect, standing out of the sea like living things, reflecting the light of heaven, or forming soft and tender shadows of the tropical sun on a blue sea. Now these same islands looked like creatures whose heads had been cut off, or like vast sarcophagi, like anything in short that reminds one of death and the grave.[75]

A landscape formed by exquisite ecological processes was abstracted and reduced to salable units. The human use of guano for agricultural enrichment likely extends back to the emerging settlement of coastal Peru 5,000 years ago.[76] By the time that Alexander von Humboldt brought samples back to France for analysis in the early 1800s, it was established as an important material for Peruvian coastal trade.[77] Shipments of Chincha Islands guano was soon headed to all corners of Europe, including Russia, Armenia, Belgium, Norway, Sweden, and France, as well as the eastern United States.[78] Of the 300,000 tons of guano exported annually from Peru in the 1860s, 40,000 of those landed in New York.[79] Indebted to Britain for aid in securing Peru's independence, Lima granted a British company control over the Peruvian guano export market. While profitable for a select Peruvian aristocracy, most of the guano profits were siphoned from the country to pay the imperial debt to Britain.[80] The arrangement functioned to bleed the country of its most valuable physical resource, with little immediate or future local investment.

Exploitation

As guano increased in value, the mountain was taken down, and it was taken down by hand. A photograph from 1865, titled "Rays of Sunlight from South America: Chinamen working guano, Great Heap, Chincha Islands," shows workers standing on ledges of the guano mountain. Like an inverted quarry, laborers picked at and shoveled the mount from the sides, slowly forming walls battered at an angle to prevent collapse. Steep steps, carved of the same guano, enabled the slow, steady removal of guano. While relatively inert as a solid material, picking at hardened guano pulverizes it, releasing a caustic cloud with each stroke. Airborne guano assaults the eyes, mouth, and lungs. Workers shoveled the loosened guano into woven bags or directly into wheelbarrows, and wheeled them to nearby docks. They tipped loads down chutes made of cane at the cliff edges, into boat holds, or otherwise carted it in bags to the launches. Direct transfer was much quicker – a hull could be filled in two days rather than two weeks. When big loads of guano were slid directly into the ships cavity, a cloud of caustic dust enveloped the boat, Peck described, "as if she were on fire with a cargo of lard."[81] Inside the ships, "trimmers" (men whose work it was to spread guano throughout the vessel) toiled in a hellish environment, where particulate guano was so thick that light could not penetrate the space. Workers wore thick bandages around their faces to protect them from ammonia and inevitable respiratory illnesses.[82]

A willing labor force for digging guano was not easy to find. Having legally abolished slavery in 1854, Peru found itself without a captive labor force to do the punishing work. Britain had abolished slavery throughout the British Empire in 1833, yet had a great stake in the profits of guano sales. To replace their newly emancipated workforce, Peru and Britain participated in a new form of slavery by hiring Chinese workers under false contracts of "free labor." Men

Figure 1.18 Original caption: "Chinamen working guano. Great Heap, Chincha Islands. Rays of Sunlight from South America."

Source: Photograph by Alexander Gardner, circa 1865 (Photography Collection, Miriam and Ira D. Wallach Division of Art, Prints and Photographs, The New York Public Library, Astor, Lenox and Tilden Foundations).

Figure 1.19 **Harvesting guano, Chincha Islands, 1862**
Source: Courtesy of New Bedford Whaling Museum.

– derogatively known as coolies – left Hong Kong and Macau as free workers (although some were kidnapped) and arrived at the Chincha Islands enslaved. From the mid-1840s through the 1870s, 100,000 men were taken from Chinese ports and sent to Peru's plantations, railroad sites, and – arguably the cruelest destination – the guano islands.[83] The *Trata Amarilla* or "yellow trade" was the direct outcome of expanding capital markets facing the end of slavery. This "debt peonage," lubricated by the growth of transportation systems, mining technology, and the global demand for agricultural and industrial minerals, took up where the Atlantic slave trade let off. And once they understood the deceptive terms and conditions of their fate, workers could

Figure 1.20 **Original caption: "Flogging a Chinese 'coolie' on board ship."**

Source: Reproduced from Edgar Holdon, "A Chapter on the Coolie Trade," in the June 1864 issue of *Harper's Magazine* by special permission.

not escape[84] – and so companies wielded slavery in a post-abolition context. Ships assumed the same violent organization of slave ships, as workers were held captive below the deck and disciplined violently. One in ten captives was said to die on the journey.[85] Those who survived the trip were sold in markets once they arrived in South American port cities. This new form of "disguised slavery," as Marx had called such systems, was integral to what Clark and Foster identify as the mid-nineteenth century's "triangle trade" of ecological imperialism: capitalist production robbed British and American soils, huge quantities of guano from Peru were extracted to "fix" them, and workers from China were "robbed" by their subjugation in the extraction.[86]

On Peck's first foray to the center island, he came across the dead body of a man, "lying almost naked in the sun, with the face of it covered with flies," near to where other workers were digging guano and amongst several small informal graves.[87] The man had drowned, but Peck suggests it could have been at his own hand. "Almost every week," he writes, "some of them commit suicide by throwing themselves from the cliff."[88] Peck estimated that there were 700 or 800 Chinese workers on the North Chincha Island in 1854,[89] and described their plight:

> They are condemned to be diggers of guano; their labor is much more severe and injurious than railroad digging; they have no liberty days, no protecting laws, no power to obtain even the pittance said to be paid them, no proper seasons of rest. Most of them go nearly naked; none have more than enough clothing just to cover themselves; they live and feed like dogs.[90]

Workers were required to gather five tons of guano a day or meet severe punishment.[91] Former slaves were forced by administrators to beat guano diggers, and these beatings frequently led to death. Torture as discipline was frequently

Figure 1.21 **Guano deposit and settlement, Chincha Islands, Peru, 1862**
Source: Courtesy of New Bedford Whaling Museum.

cited: men were tied to buoys in the ocean, chained to rocks, and hung from ship masts.[92] A "collection of hovels" as Peck put it, included a cane-shanty hospital, groceries, a cookery, and carpenter-shop, with overseers, drivers, doctors, "and nondescripts in ragged ponchos."[93] Now, on the island, the only evidence of its former inhabitants that I can see is a closed-down hospital and a vacant hotel, both bombarded by aerial guano. They appear as if they are from the twentieth century, but Jhuneor doesn't know when they were built. Jhuneor points to a distant peak on the island, and tells me it is the old cemetery for guano workers, explaining, "a lot of people died here."

As Peck left the Chincha Islands, he reflected:

> The islands themselves are in the highest degree wonderful and picturesque; the formation of the guano is a fruitful subject for curious speculation, but it was dusty, and there was something too much of it; and the Coolies and their fate impart a gloomy coloring to the recollection of the whole time spent there.[94]

Birds too, according to Peck, absorbed the melancholy of the horrendous crimes happening on the islands. To him, their "unearthly" song sounded almost human, warning visitors to leave: "'Wah, wah,' they say, 'what do you do here? Leave this place – leave this place. It is wild and sad – wah, wah. It is ours; leave it. Fly away – away. Wah, wah, wah!'"[95] Alongside the razing of guano, bird populations suffered. Unaccustomed to predators, birds were easily slaughtered for food and sport, and their habitats destroyed by the invasive harvesting process. Near the end of the guano boom on the Chincha Islands, bird populations were already scarce.[96]

That this particular history unfolded on islands is no coincidence. Surrounded by tumultuous waters, the Chincha Islands were impossible to flee. Offshore and far from societal gaze, guano companies perpetuated slavery under the auspices of countries that had already outlawed it. "It is moreover, the worst and most cruel slavery of all forms of slavery that exist among civilized nations," Peck wrote.

> It was universally said to be so by captains who had visited every quarter of the globe; I am sure I never saw anything like it on the plantations, when I once resided for several months in the country adjoining that where the scene of the popular tale of *Uncle Tom's Cabin* is laid.[97]

While Peck and others reported on the atrocities, the trade continued uncontested.[98]

As guano fever raged in urbanizing American and European centers, appetites could not be satisfied. Markets of doctored and fake guano prompted its regulation and inspection by New York authorities,[99] and guano harvest expanded to different regions around the world. Peru maintained control of its superior market, outraging the United States with a perceived "monopoly." The United States responded with the Guano Islands Act of 1856, which gave citizens the right to seize uninhabited islands containing guano deposits. The act also allowed its users to harvest guano through a monopoly, excluding all foreign vessels, and without future responsibility for the islands. More than sixty islands in the Pacific Islands were claimed, as well as forty others in the Atlantic and Caribbean.[100]

The American Guano Company claimed Baker Island in 1857. Eighty to 100 men from Sandwich Island worked the one-mile wide atoll on a seasonal basis, launching upwards of a hundred tons daily towards American ports in American boats.[101] From these shipments came the "American" guano that was mixed in with Peruvian guano for the Central Park guano mix mentioned earlier. Baker Island guano was perceived to be less potent than Peruvian, though the combination was a good compromise: a cheaper solution than Peruvian guano alone, still effective, and involving material purchased from an American venture. Baker Island deposits were largely emptied out within thirty years, and eventually placed under the jurisdiction of the U.S. Secretary of the Interior in 1936.[102] While there is little documentation about the extent of guano use in Central Park after the original construction, guano's use in American agriculture came to an end almost as rapidly as it began.

The Lawn-Industrial Complex

Peru's guano age ended abruptly with the Pacific War of 1879, in which Chile took control of Peru and Bolivia's reserves, steering the world's fertilizer markets towards nitrates or saltpeter, a more abundant nutrient source. Foreseeing the inevitable depletion of nitrates, chemists and capitalists sought synthetic alternatives. By early in the twentieth century, Fritz Haber had successfully synthesized ammonia from atmospheric nitrogen, and Carl Bosch had formulated it as a commercial product; the so-called Haber-Bosch process would change the course of global agriculture. While in the early 1900s, 90 percent of the U.S. fertilizer nitrogen came from organic materials and manures (including guano), as the chemical-fertilizer burgeoned by mid-century, only 4 percent did.[103] A true industrial-complex had developed around industrialized agriculture, offering fertilizers, fungicide, and pesticides derived of fossil-fuel feedstocks from other industries, such as coal gas from coke production. Seeking markets to utilize surplus stock and expand their reach, companies set their sights on the farm's symbolic offspring: the American lawn.[104]

Lawn-care vendors focused on new suburban yards, specialized sports fields, golf courses, and municipal landscaping as bounteous new surfaces on which to wield their wares. Monsanto, DuPont, and other companies marketed the dream of the perfect carpet lawn through "time-saving" lawn-care systems that included mixed nitrogen-phosphorus-potassium (N-P-K) fertilizers, engineered seed, pesticides, and equipment.[105] A lush, homogeneous lawn soon became the underlay of the American dream. The greener, more neatly

Figure 1.22 **Sheep Meadow, Central Park, 2013**

Source: Photograph by Jonathan Ellgen.

shorn, and more homogeneous the surface, the better. Chemically fertilized turf grass now pervades the American landscape, from golf courses to Central Park's Sheep Meadow, covering an estimated 28 million acres.[106] The impacts are enormous: one-third of all potable water and 70 million tons of pesticides are drained annually into lawn care.[107] One of the more banal materials in the landscape designer's palette, turf grass, is, for many, their bread and butter, as lawn care amounts to a $30 billion industry.[108] Frederick Law Olmsted's directives for Central Park's meadow turf quoted earlier, "The surface . . . can hardly be made too fine or too smooth and even," are realized a century later in a hyperactive fashion.

Chemical fertilizers flood the nation's turf and drain into its rivers. Runaway nitrates eventually end up in water bodies, feeding algae and water plants, creating a glut of growth, depleting oxygen and killing off other aquatic life. Invisible and distributed throughout America's most favored landscapes, runoff pollution from lawn fertilizers easily skirts regulation. Because fertilizer runoff does its damage far from the site of application, it is hard to regulate and also hard to elicit consumer concerns about its consequences.

In Central Park, fertilizer use and grass maintenance has changed over the past century and a half. Parks maintenance practices kept pace with new products and methods from the lawn-care industry and included some of their own experimental innovations. During the 1930s, the Parks Department established a 27-acre turf laboratory in Pelham Bay Park to test organic and inorganic fertilizer applications, soil mixes, and grass seed for its 10,000 acres of turf. An in-house grass scientist and soil experts developed different grass types for different conditions, including the sod cut and installed in the baseball diamonds in Central Park.[109] To this day, chemical fertilizers continue to be applied throughout the park; however, there has recently been a growing interest in lower-input fertilization techniques, including the composting of the park's own waste. But rather than horse manure and night soil, park gardeners now compost leaves and woody matter. "Compost tea" is brewed from the heap, mixed with starches, diluted, and applied to the Park's largest lawns twice yearly.[110] As well, a type of modern night soil has returned to the park: biosolids – processed sewage sludge – from the Merrimack Wastewater Treatment Facility in New Hampshire were used in a recent restoration of the Great Lawn in Central Park.[111] Algal blooms in the park's water bodies due to runoff nitrogen and phosphorus from within and beyond the park's boundaries are a recurring problem; however, Tina M. Nelson, coordinator of the Soil, Water, and Ecology Laboratory at the Central Park Conservancy, is opposed to chemical "fixes" and treats them instead with aeration.[112] The laboratory monitors the Park's soil chemistry, water quality, and animal populations in order to address landscape issues more precisely, using organic and low-input techniques. Perhaps the sheep, removed from the Sheep Meadow in the 1930s, will graze again.

Three decades after the guano industry's collapse at the end of the 1870s, *La Compania Administradora del Guano* in Peru was established to nationalize and produce a government-run sustainable industry. Guano harvesting has continued to a lesser degree throughout the century, devastated by the decimation of the bird population[113] due to overfishing by industrial-scale anchovy fisheries (fueling another fertilizer industry: fishmeal) and exacerbated by the effects of El Niños in 1965 and 1972.[114] The Guano Islands and Capes National Reserve was launched in 2009 to legalize and formalize the conservation area, to preserve the islands' marine and land biodiversity, and, secondarily, to support the regulated sustainable harvest of guano from the system's twenty-two islands and eleven capes.

Harvesting occurs "selectively" by rotating between islands once guano has accumulated sufficiently; as well, commercial fisheries are banned from the area, and artisanal fishing operations are encouraged. The guano is sold to small farmers for a primarily domestic market. Quechua-speaking workers from the Andes, who can earn more collecting guano than in the impoverished highlands, migrate to the islands seasonally to harvest. Chilean photographer Tomàs Munita has documented the contemporary seasonal harvest on North Guañape Island (north of Lima) in his series, "Islands of Sorrow."[115] His photographs are eerily out of time; the guano dust casts a sepia hue on the men who use similar tools and techniques as guano harvesters did 150 years ago.[116]

At the Guano Islands and Capes National Reserve's on-land office in Paracas, Jhuneor and his colleagues tell me about the aims and challenges of the

Figure 1.23 **Guano harvesting, North Guañape Island, Peru, 2009**
Source: Photograph by Tomas Muñita.

Figure 1.24 Jhuneor Paitan Ñahui, Rodrigo
Ramirez, and colleague, SERNANP Office, Paracas,
Peru, 2014

program. It has gained recognition as an innovative sustainable management
program, yet the scope of their work is huge and funds are scarce; for example,
they really need a boat to do their work. Jhuneor sees carefully managed tourism, like that of the nearby Ballestas Islands, as a key to generating financial
support for their work. The Ballestas are often called the "poor man's Galapagos," referring to the fact that they contain remarkable biodiversity, and
are relatively cheap and easy to visit and, therefore, more accessible than the
Galapagos' luxury tourism industry. Tourism, Jhuneor remarks, if not properly
managed, can jeopardize the very conditions that it might support.

Overdraft

Guided by the logic of growth, capitalists viewed the Chincha guano as an
inert stockpile of potential – both in terms of capital "growth" and because
this material would support agricultural plant growth in the near future. From
their perspective, merchants and their governmental associates activated this
stock, distributing it so that it could be useful and accelerate life in distant
places; through extraction and transport, they were agents of nutrient circulation, connecting guano to their own depleted soils. The decimation of the Chincha Islands guano landscape and bird populations, as well as the oppression
of enslaved workers, were unseen consequences of agriculture's modernization
and the urban growth it supported.

When American and European farms imported Peruvian guano to replenish and charge their exhausted soils, a massive displacement of nutrients
occurred at a global scale. Clark and Foster use the term "environmental over-
draft" to describe one nation or power's act of drawing on other territories in
order to benefit their own, the nineteenth-century guano trade being a prime
example. This episode of ecological imperialism transformed landscapes in different places and in radically different ways. In Central Park and the New York
area farms that the park was a metaphor for, guano improved soils as had

never been seen before, offering better yields, preferred qualities, and, most significantly, new expectations for growth. Guano's disappearance stimulated the development of the contemporary chemical fertilizer industry, in both agriculture and lawn care, as a central facet of the public landscape. The highly toxic chemicals behind America's lush lawns are a well-recognized contradiction of the modern landscape. In the Chincha Islands, the guano trade transformed the landscape by physically removing most of it and, through the massacre of bird populations, disrupted its ability to function as it had. It was used to innovate and justify a Pacific slavery, and involved economic arrangements that benefited Britain at Peru's long-term expense. Today in the Chincha Islands, conservation initiatives are challenged by the legacy of the country's ongoing poverty, with tourism offering one of few ways forward. Ecotourism in Peru and the lurid, high-input modern lawn (as well as the movements to "improve" it) in the United States are the culmination of this material exchange. They are both descendants of idealized landscapes with contradictory pasts – contradictions ripe for unpacking.

Notes

1   Frederick Law Olmsted, "Difficulties of Preserving Green Turf, 15 May 1874," in *Forty Years of Landscape Architecture*, eds. Frederick Law Olmsted Jr. and Theodora Kimball (Cambridge, MA: MIT Press, 1973), 427.
2   Frederick Law Olmsted, "Superintendent of Central Park to Gardeners, 1873," in *Forty Years of Landscape Architecture: Central Park [by] Frederick Law Olmsted, Sr.*, ed. Frederick Law Olmsted Jr. and Theodora Kimball (Cambridge, MA: MIT Press, 1973), 358.
3   While landscape architect Frederick Law Olmsted and architect Calvert Vaux both authored the Greensward Plan, the winning competition entry for the design of Central Park, Olmsted oversaw matters most related to the growth of plants, so I'll primarily refer to him in this chapter.
4   Frederick Law Olmsted and Calvert Vaux, "A Review of Recent Changes, and Changes which Have Been Projected, in the Plans of the Central Park, 1872," in *Forty Years of Landscape Architecture*, eds. Frederick Law Olmsted Jr. and Theodora Kimball (Cambridge, MA: MIT Press, 1973), 249.
5   Frederick Law Olmsted, "Letter of Frederick Law Olmsted to the Commissioners of Central Park, 31 May 1858," in *Forty Years of Landscape Architecture*, 46.
6   Ibid.
7   Ibid., 243.
8   "Central Park Sheep: A Flock of Sixty Dorsets and Four Brazilians Kept in the West Side Meadow," *New York Tribune*, June 2, 1901, B6.
9   "The City's Sheep Flock: It Is a Benefit to Central Park as Well as a Curiosity and an Ornament," *New York Times*, December 17, 1899, 11.
10 "Central Park Sheep," B6.
11 "The City's Sheep Flock," 11.
12 Board of Commissioners of the Central Park, *Fifth Annual Report of the Board of Commissioners of the Central Park* (New York: Wm. C. Bryant & Co., 1862), 120.
13 Ibid.
14 Ibid.
15 Ibid., 121.
16 Board of Commissioners of the Central Park, *Seventh Annual Report of the Board of Commissioners of the Central Park* (New York: Wm. C. Bryant & Co., 1864), 87.

17 Olmsted, "Superintendent of Central Park to Gardeners, 1873," 357–8.

18 Roy Rosenzweig and Elizabeth Blackmar, *The Park and the People: A History of Central Park* (Ithaca: Cornell University Press, 1989), 246.

19 Ibid. Alvaro Sevilla-Buitrago examines the ways in which Central Park functioned as an enclosure regime designed to limit the spontaneous and disruptive tendencies of the working class to organize in public space; see Alvaro Sevilla-Buitrago, "Central Park against the Streets: The Enclosure of Public Space Cultures in Mid-Nineteenth Century New York," *Social & Cultural Geography* 15, no. 2 (2014): 151–71.

20 Rosenzsweig and Blackmar, *The Park and the People*, 88.

21 Ibid., 70.

22 Frederick Law Olmsted and Calvert Vaux, "Particulars of Construction and Estimate, 1858," in *Forty Years of Landscape Architecture*, 284.

23 Joel A. Tarr and Clay McShane, "The Centrality of the Horse to the Nineteenth-Century American City," in *The Making of Urban America*, ed. Raymond A. Mohl (New York: Rowman & Littlefield, 1997), 121.

24 Ibid.

25 George Edwin Waring, "The Disposal of a City's Waste," *The North American Review* 161, no. 464 (1895): 55.

26 Rosenzsweig and Blackmar, *The Park and the People*, 222.

27 Board of Commissioners of the Central Park, *Thirteenth Annual Report of the Board of Commissioners of the Central Park, for the Year Ending December 31, 1869* (New York: Evening Post Steam Presses, 1870), 62.

28 "Fertilizing the Park Lawns," *New York Tribune*, February 15, 1885, 4.

29 Joanne Abel Goldman, *Building New York's Sewers: Developing Mechanisms of Urban Management* (Indiana: Purdue University Press, 1997), 25.

30 Peter C. Baldwin, *In the Watches of the Night: Life in the Nocturnal City, 1820–1930* (Chicago: University of Chicago Press, 2012), 107.

31 Board of Commissioners of the Central Park, *Fifth Annual Report of the Board of Commissioners of the Central Park*, 120.

32 The sewage system developed slowly until Senator William Tweed's 1870 overhaul of the Department of Public Works; see Goldman, *Building New York's Sewers*, 146.

33 Ibid., 10.

34 Egbert Ludovicus Viele, *Report on Civil Cleanliness, and the Economical Disposition of the Refuse of Cities* (New York: Edmund Jones & Co., Printers, 1860), 33.

35 George E. Waring, *Earth-Closets: How to Make Them and How to Use Them* (New York: Tribune Association, 1868), 3.

36 Ibid., 26.

37 Richard A. Wines, *Fertilizer in America: From Waste Recycling to Resource Exploitation* (Philadelphia: Temple University Press, 1985), 11.

38 "Manures: Nature's Reciprocity System," *The Southern Cultivator* 4 (1846): 167.

39 Joel A. Tarr, "From City to Farm: Urban Wastes and the American Farmer," *Agricultural History* 49, no. 4 (1975): 604.

40 Wines, *Fertilizer in America*, 12.

41 Ibid., 26–7.

42 Ibid., 25.

43 Ibid., 37.

44 Charles L. Bartlett, *Guano: A Treatise on the History, Economy as a Manure, and Modes of Applying Peruvian Guano, in the Culture of the Various Crops of the Farm and the Garden* (Boston: C.L. Bartlett, 1860), 7.

45 Ibid., 8–9.

46 Wines, *Fertilizer in America*, 38.

47 Gregory T. Cushman, *Guano and the Opening of the Pacific World: A Global Ecological History* (New York: Cambridge University Press, 2014), 44; Wines, *Fertilizer in America*, 39.

48 Letter from Justus Von Liebig to John Mechi, November 17, 1859, reprinted in "On English Farming and Sewers," *Monthly Review*, July 1, 2018, https://monthly review.org/2018/07/01/on-english-farming-and-sewers.

49 Karl Marx, *Capital, Vol. 1* (London: Penguin, 1976), 637–8.

50 John Bellamy Foster, "Marx's Theory of Metabolic Rift," *American Journal of Sociology* 105 no. 2 (1999): 384.

51 Brett Clark and John Bellamy Foster, "Ecological Imperialism and the Global Metabolic Rift: Unequal Exchange and the Guano/Nitrates Trade," *International Journal of Comparative Sociology* 50, no. 3–4 (2009): 312–13.

52 Frederick Law Olmsted, *A Journey in the Seaboard Slave States: With Remarks on Their Economy* (New York: Dix & Edwards; London: Sampson Law, Son & Co., 1856), 752. I was alerted to Olmsted's discussion of guano by Jennifer C. James in "Buried in Guano: Race, Labor, and Sustainability," *American Literary History* 24, no. 1 (2012): 115–42.

53 Rosser H. Taylor, "Fertilizers and Farming in the Southeast, 1840–1950, Part I: 1840–1900," *The North Carolina Historical Review* 30, no. 3 (1953): 306.

54 Ibid., 310.

55 Olmsted, *A Journey in the Seaboard Slave States*, 9.

56 Ibid., 303.

57 James, "Buried in Guano," 124.

58 Olmsted, *A Journey in the Seaboard Slave States*, 105.

59 Ibid.

60 Taylor, "Fertilizers and Farming in the Southeast, 1840–1950, Part I: 1840–1900," 310.

61 James, "Buried in Guano," 125.

62 Taylor, "Fertilizers and Farming in the Southeast, 1840–1950, Part I: 1840–1900," 312.

63 Cushman, *Guano and the Opening of the Pacific World*, 85.

64 "New Slave Trade Frightful Revelation," *Frederick Douglass' Paper*, December 14, 1855.

65 Servicio Nacionales de Areas Protegidas (SERNANP), "Sistema de Islas, Islotes y Puntas Guaneras," accessed January 21, 2019, www.sernanp.gob.pe/sistema-de-islas-islotes-y-puntas-guaneras; Jhuneor Paitan Ñahui, personal communication, August 24, 2014.

66 George Washington Peck, *Melbourne, and the Chincha Islands: With Sketches of Lima, and a Voyage Round the World* (New York: Charles Scribner, 1854), 168.

67 Cushman, *Guano and the Opening of the Pacific World*, 5.

68 Peck, *Melbourne, and the Chincha Islands*, 174.

69 Jaime Jahncke, David M. Checkley Jr., George L. Hunt Jr., "Trends in Carbon Flux to Seabirds in the Peruvian Upwelling System: Effects of Wind and Fisheries on Population Regulation," *Fisheries Oceanography* 13, no. 3 (2004): 209.

70 Peck, *Melbourne, and the Chincha Islands*, 169.

71 Bartlett, *Guano*, 5.

72 George Evelyn Hutchinson, "The Biogeochemistry of Vertebrate Excretion," *Bulletin of the American Museum of Natural History* 96 (1950): 66, 70.

73 Ibid., 70.

74 Wines, *Fertilizer in America*, 45.

75 Alexander James Duffield, *Peru in the Guano Age: Being a Short Account of a Recent Visit to the Guano Deposits, With Some Reflections on the Money They Have Produced and the Uses to Which It Has Been Applied* (London: R. Bentley and Son, 1877), 89.

76 Cushman, *Guano and the Opening of the Pacific World*, 8.

77 Ibid., 25.

78 Clark and Foster, "Ecological Imperialism and the Global Metabolic Rift," 322.

79 "Interesting to Farmers: Guano Its History, Traffic, Uses, Abuses, and Frauds Price of Peruvian Guano," *New-York Daily Tribune*, September 29, 1860, 8.

80 Clark and Foster, "Ecological Imperialism and the Global Metabolic Rift," 319–21.

81 Peck, *Melbourne, and the Chincha Islands*, 209–11.

82 Ibid., 211.

83 Edward D. Melillo, "The First Green Revolution: Debt Peonage and the Making of the Nitrogen Fertilizer Trade, 1840–1930," *American Historical Review* 117, no. 4 (2012): 1029.

84 Ibid., 1030, 1031.

85 Ibid., 1039.

86 Clark and Foster, "Ecological Imperialism and the Global Metabolic Rift," 330.

87 Peck, *Melbourne, and the Chincha Islands*, 165.

88 Ibid., 208.

89 Ibid., 206.

90 Ibid., 207.

91 Ibid., 208.

92 David Hollett, *More Precious Than Gold: The Story of the Peruvian Guano Trade* (Madison: Fairleigh Dickinson University Press, 2008), 130–1.

93 Peck, *Melbourne, and the Chincha Islands*, 208.

94 Ibid., 214.

95 Ibid., 167.

96 Hutchinson, "The Biogeochemistry of Vertebrate Excretion," 51.

97 Peck, *Melbourne, and the Chincha Islands*, 212.

98 Clark and Foster, "Ecological Imperialism and the Global Metabolic Rift," 324.

99 Jimmy M. Skaggs, *The Great Guano Rush: Entrepreneurs and American Overseas Expansion* (New York: Palgrave Macmillan, 1994), 2.

100 Dan O'Donnell, "The Pacific Guano Islands: The Stirring of American Empire in the Pacific Ocean," *Pacific Studies* 16, no. 1 (1993): 43.

101 "Life on a Guano Island," *New York Times*, September 9, 1866, 2.

102 O'Donnell, "The Pacific Guano Islands," 57.

103 Kristoffer Whitney, "Living Lawns, Dying Waters," *Technology & Culture* 51, no. 3 (2010): 662.

104 Ibid., 664.

105 Ibid., 665.

106 U.S. Environmental Protection Agency, "Landscape Fact Sheet," accessed June 20, 2015, www.epa.gov/region1/lab/pdfs/LabLandscapeFactsheet.pdf.

107 Ibid.

108 Ibid.

109 Sanderson Vanderbilt, "Park Grass Is Growing Greener as Science Replaces Guessing," *New York Herald Tribune*, November 25, 1934, A3.

110 Lisa W. Foderaro, "From Lawn to Pond, a Guardian of Central Park," *The New York Times*, February 21, 2012, www.nytimes.com/2012/02/22/nyregion/in-a-central-park-laboratory-providing-the-diagnosis-for-an-843-acre-patient.html.

111 New England Biosolids & Residuals Association, "Merrimack Biosolids: Nourishing Greener Parks and Fairways," *NEBRA Biosolids Case Study 5*, June 2001, www.biosolids.com/Headlines/pdf/NEBRACaseStudyMerrimack01.pdf.

112 Foderaro, "From Lawn to Pond."

113 Simon Romero, "Peru Guards Its Guano as Demand Soars Again," *The New York Times*, May 30, 2008, www.nytimes.com/2008/05/30/world/americas/30peru.html.

114 Jahncke, Checkley Jr., and Hunt Jr., "Trends in Carbon Flux to Seabirds in the Peruvian Upwelling System," 209.

115 Tomàs Munita, "Island of Sorrow," Peru, 2008, www.tomasmunita.com/guano.

116 Romero, "Peru Guards Its Guano as Demand Soars Again".

Bibliography

Baldwin, Peter C. *In the Watches of the Night: Life in the Nocturnal City, 1820–1930.* Chicago: University of Chicago Press, 2012.

Bartlett, Charles L. *Guano. A Treatise on the History, Economy as a Manure, and Modes of Applying Peruvian Guano, in the Culture of the Various Crops of the Farm and the Garden.* Boston: C.L. Bartlett, 1860.

Board of Commissioners of the Central Park. *Fifth Annual Report of the Board of Commissioners of the Central Park.* New York: Wm. C. Bryant & Co., 1862.

Board of Commissioners of the Central Park. *Seventh Annual Report of the Board of Commissioners of the Central Park.* New York: Wm. C. Bryant & Co., 1864.

Board of Commissioners of the Central Park. *Thirteenth Annual Report of the Board of Commissioners of the Central Park, for the Year Ending December 31, 1869.* New York: Evening Post Steam Presses, 1870.

"Central Park Sheep: A Flock of Sixty Dorsets and Four Brazilians Kept in the West Side Meadow." *New York Tribune,* June 2, 1901.

"The City's Sheep Flock: It is a Benefit to Central Park as Well as a Curiosity and an Ornament." *New York Times,* December 17, 1899.

Clark, Brett, and John Bellamy Foster. "Ecological Imperialism and the Global Metabolic Rift: Unequal Exchange and the Guano/Nitrates Trade." *International Journal of Comparative Sociology* 50, nos. 3–4 (2009): 311–34.

Cushman, Gregory T. *Guano and the Opening of the Pacific World: A Global Ecological History.* New York: Cambridge University Press, 2014.

Duffield, Alexander James. *Peru in the Guano Age: Being a Short Account of a Recent Visit to the Guano Deposits, With Some Reflections on the Money They Have Produced and the Uses to Which It Has Been Applied.* London: R. Bentley and Son, 1877.

"Fertilizing the Park Lawns." *New York Tribune,* February 15, 1885.

Foderaro, Lisa W. "In a Central Park Laboratory, Providing the Diagnosis for an 843-Acre Patient." *New York Times,* February 21, 2012, www.nytimes.com/2012/02/22/nyregion/in-a-central-park-laboratory-providing-the-diagnosis-for-an-843-acre-patient.html.

Foster, John Bellamy. *The Ecological Revolution: Making Peace with the Planet.* New York: Monthly Review Press, 2009.

Goldman, Joanne Abel. *Building New York's Sewers: Developing Mechanisms of Urban Management.* West Lafayette: Purdue University Press, 1997.

Hollett, David. *More Precious than Gold: The Story of the Peruvian Guano Trade.* Madison: Fairleigh Dickinson University Press, 2008.

Hutchinson, George Evelyn. "The Biogeochemistry of Vertebrate Excretion." *Bulletin of the American Museum of Natural History* 96 (1950).

"Interesting to Farmers: Guano Its History, Traffic, Uses, Abuses, and Frauds Price of Peruvian Guano." *New-York Daily Tribune,* September 29, 1860.

Jahncke, Jaime, David M. Checkley Jr., and George L. Hunt Jr. "Trends in Carbon Flux to Seabirds in the Peruvian Upwelling System: Effects of Wind and Fisheries on Population Regulation." *Fisheries Oceanography* 13, no. 3 (2004): 208–23.

James, Jennifer C. "Buried in Guano: Race, Labor, and Sustainability." *American Literary History* 24, no. 1 (2012): 115–42.

Liebig, Justus von. "Letter to John Mechi, 17 November 1859, reprinted in 'On English Farming and Sewers'." *Monthly Review,* July 1, 2018, https://monthlyreview.org/2018/07/01/on-english-farming-and-sewers.

"Life on a Guano Island." *New York Times,* September 9, 1866.

"Manures: Nature's Reciprocity System." *The Southern Cultivator* 4 (1846): 167.

Melillo, Edward D. "The First Green Revolution: Debt Peonage and the Making of the Nitrogen Fertilizer Trade, 1840–1930." *American Historical Review* 117, no. 4 (2012): 1028–60.

Ñahui, Jhuneor Paitan. Personal communication, August 24, 2014.

New England Biosolids & Residuals Association. "Merrimack Biosolids: Nourishing Greener Parks and Fairways." *NEBRA Biosolids Case Study 5*, June 2001. Accessed June 18, 2015, www.biosolids.com/Headlines/pdf/NEBRACaseStudyMerrimack01. pdf.

"New Slave Trade Frightful Revelation." *Frederick Douglass' Paper*, December 14, 1855.

O'Donnell, Dan. "The Pacific Guano Islands: The Stirring of American Empire in the Pacific Ocean." *Pacific Studies* 16, no. 1 (1993): 43–66.

Olmsted, Frederick Law. "Difficulties of Preserving Green Turf, 15 May 1874." In *Forty Years of Landscape Architecture*, edited by Frederick Law Olmsted Jr. and Theodora Kimball, Cambridge, MA: MIT Press, 1973.

Olmsted, Frederick Law. *A Journey in the Seaboard Slave States: With Remarks on Their Economy*. New York: Dix & Edwards, 1856.

Olmsted, Frederick Law. "Letter of Frederick Law Olmsted to the Commissioners of Central Park, 31 May 1858." In *Forty Years of Landscape Architecture: Central Park [by] Frederick Law Olmsted, Sr.*, edited by Frederick Law Olmsted Jr. and Theodora Kimball, Cambridge, MA: MIT Press, 1973.

Olmsted, Frederick Law. "Superintendent of Central Park to Gardeners, 1873." In *Forty Years of Landscape Architecture*, edited by Frederick Law Olmsted Jr. and Theodora Kimball, Cambridge, MA: MIT Press, 1973.

Olmsted, Frederick Law, and Calvert Vaux. "Particulars of Construction and Estimate, 1858." In *Forty Years of Landscape Architecture*, edited by Frederick Law Olmsted Jr. and Theodora Kimball, Cambridge, MA: MIT Press, 1973.

Olmsted, Frederick Law, and Calvert Vaux. "A Review of Recent Changes, and Changes which Have Been Projected, in the Plans of the Central Park, 1872." In *Forty Years of Landscape Architecture*, edited by Frederick Law Olmsted Jr. and Theodora Kimball, Cambridge, MA: MIT Press, 1973.

Peck, George Washington. *Melbourne, and the Chincha Islands: With Sketches of Lima, and a Voyage Round the World*. New York: Charles Scribner, 1854.

Romero, Simon. "Peru Guards Its Guano as Demand Soars Again." *The New York Times*, May 30, 2008, www.nytimes.com/2008/05/30/world/americas/30peru.html.

Rosenzsweig, Roy, and Elizabeth Blackmar. *The Park and the People: A History of Central Park*. Ithaca: Cornell University Press, 1992.

Servicio Nacionales de Areas Protegidas (SERNANP). "Sistema de Islas, Islotes y Puntas Guaneras." Accessed January 21, 2019, www.sernanp.gob.pe/sistema-de-islas-islotes-y-puntas-guaneras.

Sevilla-Buitrago, Alvaro. "Central Park Against the Streets: The Enclosure of Public Space Cultures in Mid-Nineteenth Century New York." *Social & Cultural Geography* 15, no. 2 (2014): 151–71.

Skaggs, Jimmy M. *The Great Guano Rush: Entrepreneurs and American Overseas Expansion*. New York: Palgrave Macmillan, 1994.

Tarr, Joel A. "From City to Farm: Urban Wastes and the American Farmer." *Agricultural History* 49, no. 4 (1975): 598–612.

Tarr, Joel A., and Clay McShane. "The Centrality of the Horse to the Nineteenth-Century American City." In *The Making of Urban America*, edited by Raymond A. Mohl. New York: Rowman & Littlefield, 1997.

Taylor, Rosser H. "Fertilizers and Farming in the Southeast, 1840–1950: Part I: 1840–1900." *The North Carolina Historical Review* 30, no. 3 (1953): 305–28.

Tomàs Munita. "Island of Sorrow." Peru, 2008. Accessed December 18, 2018, www.tomasmunita.com/guano.

U.S. Environmental Protection Agency. "Landscape Fact Sheet." Accessed June 20, 2015, www.epa.gov/region1/lab/pdfs/LabLandscapeFactsheet.pdf.

Vanderbilt, Sanderson. "Park Grass Is Growing Greener as Science Replaces Guessing." *New York Herald Tribune*, November 25, 1934.

Viele, Egbert Ludovicus. *Report on Civil Cleanliness, and the Economical Disposition of the Refuse of Cities*. New York: Edmund Jones & Co., Printers, 1860.

Waring, George Edwin. "The Disposal of a City's Waste." *The North American Review* 161, no. 464 (1895): 49–56.

Waring, George Edwin. *Earth-Closets: How to Make Them and How to Use Them*. New York: Tribune Association, 1868.

Whitney, Kristoffer. "Living Lawns, Dying Waters." *Technology & Culture* 51, no. 3 (2010): 652–74.

Wines, Richard A. *Fertilizer in America: From Waste Recycling to Resource Exploitation*. Philadelphia: Temple University Press, 1985.

Figure 2.1(A) Detail, Sands Quarry in
Vinalhaven, Maine, 1907

Source: Photograph by T.N. Dale (U.S. Geological
Survey Bulletin 313: 1907. U.S. Geological Survey
Department of the Interior/USGS).

— to Broadway, 1892

Figure 2.1(B) Detail, paving, and construction of cable car line on Broadway, 1891

Source: Photograph by C.C. Langill and William Gray (Photography Collection, Miriam and Ira D. Wallach Division of Art, Prints and Photographs, The New York Public Library, Astor, Lenox and Tilden Foundations).

Chapter 2

Range of Motions

Granite from Vinalhaven, Maine, to Broadway, 1892

Granite moves first as molten material, slowly cooling to solid rock. As it passes through different hands, tools, and landscapes, it changes form, use, and value. A "motion," in late-nineteenth-century quarrying terminology, is the work area drawn by the rotation of a quarry derrick, but it is also a specific type of small quarry where paving blocks are cut directly from the ground. Quarry workers and paving cutters in New England quarries transformed the bedrock of their islands into the paving blocks that were sent on ships to harden the streets of urbanizing eastern cities. Smooth, granite-paved streets in New York City in the late nineteenth century, such as Broadway, lubricated the flow of capital and the rapid construction of buildings and landscapes alongside it. Stone surfaces reduced dust and mud, allowing for greater speeds and efficient transactions. At the same time, New York's appetite for granite structured social life and labor movements in Maine. Large urban contracts for paving and building blocks supported the expansion of granite-producing towns like Vinalhaven, the largest of the Fox Islands in the Penobscot Bay, Maine.

The concept of "material flow" conjures flowcharts with neutral arrows darting around the page, indicating the unidirectional movement of abstract material streams between extraction and use sites. "Flow" on its own implies easy and unencumbered movement. Such terms help conceptualize the circulation and spatial extents of industry and resource extraction in a world of globalized capital, yet they obscure the physicality of the material itself and also tend to conceal the role of human labor in facilitating the movement of matter. This chapter traces the passage of granite paving blocks from Vinalhaven, Maine, to Broadway, New York City, examining the ways in which human work and granite's very properties enabled that flow.

Rock Motion

When molten magma pierces the earth's surface, it cools rapidly. Quenched by the cool air or the ocean's water, it quickly hardens into glassy or fine-grained rocks like obsidian or basalt. But when magma is contained by overlying rock and pressured from below, it cools more slowly. In the cooling process, the elements in the magma have time to crystalize, grow, and co-mingle as distinct mineral grains. This slow organization and cooling under pressure makes the coarse-grained rock that we recognize as "granite." A fluid mass transforms into an orderly assembly of solid, differentiated minerals, each solidifying in a sequence according to their unique crystallization temperatures. First, biotite and hornblende crystallize at higher temperatures, and then as the mass cools closer to the earth's surface, mica, feldspars, and quartz, which crystallize at lower temperatures, begin to solidify.[1] While steadfast and silent, granite's patterns and textures betray its former movement. Large grains indicate a slower cooling speed in which minerals had time to grow; small, speckled grains, the size of sand, reflect a more rapid cooling time where the formation of minerals had only just begun. In the early twentieth century, geologist T. Nelson Dale noticed that, even without a microscope, one could see how individual minerals interfere with each other's formation; the rock's surface is a diagram of the minerals' interactions.[2] The orientation of certain minerals also revealed something of the magma's former motion. Dale described the "flow structure" of granites in various quarries of New England – how bands of mica and other minerals trace the direction that the molten mass once flowed under pressure.[3]

While some geologists might ponder the formative processes of rock, architects consider durability in terms of structure and cultural taste. While

Figure 2.2 **Mineralogical analysis of granite from Nelson's Quarry, Vinalhaven**

Source: Courtesy of Vinalhaven Historical Society.

some might examine the surrounding landscape to understand intrusion, others examine it to locate and estimate financial returns in relation to transport and market. While geological classification is based on mineralogical composition, economical classification is based on desirable surface characteristics fit for commercial production. Some look back in time, others forward. Translating between geology and industry, T. Nelson Dale wrote the definitive *Granites of Maine* in 1907 for geology students and architects alike.[4] Any quarryman would eventually need to understand the geological underpinnings of their trade, and many aspects of geological interest translated directly into issues of economic importance and architectural applications. For example, a geological understanding of how joints and imperfections function in the stone is imperative for quarry management. An understanding of how sap – the moisture found in quarry rock – works helps the quarry superintendent to cut along it to prevent rust-colored discoloration due to oxidizing minerals. At the same time, "pleasing contrasts" could be produced through the use of the discolored stone with light-colored granites in construction.[5] Aesthetic properties are formed at the scale of the mineral. For example, the presence and distribution of different feldspars determine color and luster, while the presence and distribution of quartz and micas determine shade.

The magma flow that solidified into current day Vinalhaven, Maine, is only a small segment of a massive belt of igneous rock. The so-called "Bays-of-Maine Igneous Complex" extends 175 miles from Maine's northern border with New Brunswick south to Penobscot Bay, where it possibly plunges beneath the surface and continues for 125 miles more, south to Massachusetts.[6] Dated to the middle Paleozoic, the complex is made of granites and gabbroic (or "black granite") rock. Like the Loch Ness monster, the complex appears and disappears; wherever granite is visible to the eye, overlying rock has eroded over millennia. All of the granites found along this belt share the same history, yet they differ depending on local conditions of overlying rock at the time of flow, and this in turn affects grain size, grain distribution, and color.

Vinalhaven is primarily composed of a pink-buff biotite granite of coarse, even-grained texture, with feldspars of up to three-quarters of an inch in diameter and quarter-inch-wide biotite flecks.[7] In an 1880 report, geologist Nathanial Shaler remarked that the stone of the region

> opens easily, having the peculiar inchoate joints that are such striking features in the syenite or granite of New England. . . . The lines of weakness in the rocks are not so numerous as to make the quarried masses too small for use.[8]

Penobscot Bay granite's workability, appealing coloration, and polish, as well as the promise of exceptionally large blocks, made it desirable for a range of applications, including for construction, sculpture, and paving. Coastal Maine's primary commercial advantage, however, was its interminable, sinuous

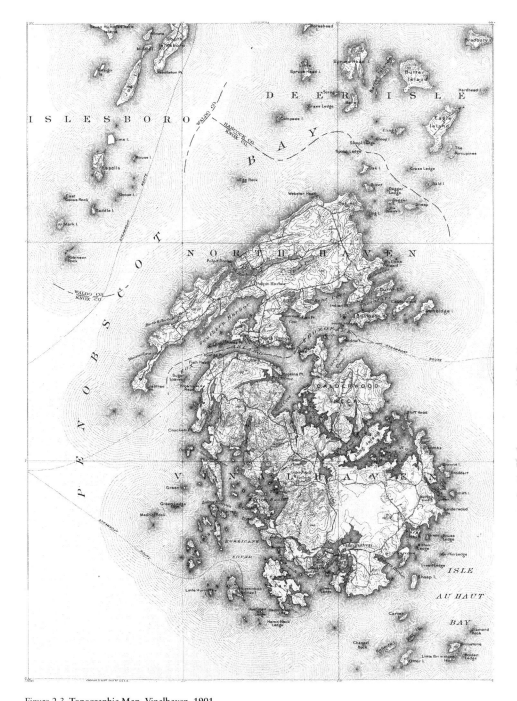

Figure 2.3 **Topographic Map, Vinalhaven, 1901**

Source: U.S. Geological Survey. Topography State of Maine, Vinalhaven Quadrangle. Reston, VA: U.S. Department of the Interior.

shoreline, where quarries were positioned so close to docks that stone could be easily loaded directly onto boats headed for Boston, New York, Philadelphia, and Washington, DC.[9]

"Where Granite Went"

Still rocking from the gusty Rockland ferry from the mainland, I disembark at Vinalhaven's central dock. I take a right onto West Main Street and a left up the hill until I spot the former Universalist Church that houses the Vinalhaven Historical Society. Inside the hall are pieces of the island's cultural history. One area showcases quilts, baskets, and furniture arranged as domestic scenes. Another area displays seines, fishing nets worn by horses and manufactured by island women. Concentrated in one corner of the museum are all things granite. A low table displays a catalogue of blocks, from the durak cube to the "Manhattan Special" and the coarse-grained "New Yorker," all surrounded by the iron hammers, wedges, and shims needed to cut and trim them. Surrounding the tools themselves is evidence of the social life that granite made: union medallions, a large embroidered tapestry from the Local Granite Cutters' union, photographs of women and men marching in parades, and Bodwell Granite Company Store receipts.

It's November, and I can see my breath in the still museum, but the Vinalhaven Historical Society office constructed inside the hall is heated and cozy. Bill Chilles and Sue Radley are working away that Saturday morning.[10] Coffee

Figure 2.4 **Paving block display, Vinalhaven Historical Society, 2013**

brews, the radio chatters, and black, three-ring binders line the bookshelves. Knowing that I've come to learn about the paving blocks that flooded north-eastern city streets, Bill and Sue have laid out a dozen or more relevant binders on the central worktable. All were intriguing, labeled with different quarry names that I had come to find out more about. But the label on one of the binders was too good to be true: "Where Granite Went." In it, page after page of photos, newspaper clippings, and bills of sale record the trajectories of paving blocks, building blocks, and carved stone from Vinalhaven. Like a scrapbook of milestones and journeys of a human lifetime, the pages show where the granite traveled to and what buildings, streets, and bridges it solidified.

Twenty-seven pages of indexed granite trips are organized alphabetically by state, starting with Connecticut (the Hartford Bridge) and ending with Wisconsin (the Solomon Jumeau Monument), dated from between the mid-nineteenth century and the early twentieth. Federal governments, banks, and new companies provided steady contracts for the island's quarries; urbanizing centers needed post offices, customs houses, banks, and federal buildings, and granite block was the most suitable material for these weighty institutions. Eagles and other ornate carved characters traveled to perch on post offices and federal buildings. Paving blocks, though less triumphant, were a formidable part of the industry, used to surface New York City, New Orleans, St. Louis, Philadelphia, and Boston. Stone destined to resist or bypass water traveled to become breakwaters, bridges, and dams. In the years immediately preceding the Civil War, civic agendas gave way to military ones, and granite moved to secure forts in Virginia, Delaware, and Florida. Immediately following the war, granite was then used to memorialize, comprising part of the Washington Monument and the 75-foot-long obelisk of the General Wool Monument in Troy, New York, in 1877, among others.

"Where Granite Went" touched on a lingering curiosity I had while on Vinalhaven: What does it feel like to send your island and your work "away"? Along with the binder, elsewhere in the museum are other displays showing the dispersion of Vinalhaven granite: one map shows how almost every state in the United States has a Vinalhaven granite horse trough. The map shows a tiny red dot on Vinalhaven, and a proliferation of red dots throughout the states – a granitic dissemination. This stone hadn't simply disappeared, leaving only holes in the ground; it was memorialized and recorded in the binder, and furthermore, it was still very sturdily somewhere. Vinalhaven's granite had transformed the urban landscapes in which it settled. It would stay there for a while, fixed in place thanks to granite's weight, durability, and esteem. It held – at least in the terms of the "Where Granite Went" binder – the status of a proud diaspora. "I'm so amazed that for such a small place, [Vinalhaven granite] is so important in New York City and Washington, DC. To have so much of us there, its crazy," Bill remarks.

He then shows me around the island in his pick-up truck. Snow hasn't fallen yet, and we can see through the bare woods as we drive. Wood buildings line the small network of streets. Paradoxically, Vinalhaven is a town constructed primarily of wood, not stone. Bill's right hand on the steering wheel

Figure 2.5 Bill Chilles and Sue Radley at work in the Vinalhaven Historical Society office, 2013

Figure 2.6 "Where Granite Went" archives, Vinalhaven Historical Society, 2013

Figure 2.7 **Paving cutter Charles Chilles**
Source: Courtesy of Vinalhaven Historical Society.

extends each time we pass another vehicle – an island salute in a small community where he seems to know everyone. One could say that Bill is on Vinalhaven because of its bedrock. "We've been in granite," he says, "forever." His great-grandfather Charles, a Scottish granite cutter, came to Vinalhaven in the mid-1800s; Charles' three brothers, also cutters, worked in quarries in Maine and Georgia. His great-uncle Fred operated one of the island's last paving block motion quarries, which closed in 1940. By the time Bill was born in the late 1960s, most of the granite operations had been long closed. He remembers hearing his grandfather tell stories of old Vinalhaven and inspecting the objects in the Vinalhaven Historical Society museum. Bill left the island briefly for school and work in Boston, but he couldn't stand the noisy city and returned for good years ago. Over the years he has become more and more involved in the historical society, and he was serving as its president when I visited.

It's possible to drive across the island in half an hour. Forty-two major quarries lined the southern half of Vinalhaven with a few tucked into inlets on the north of the island. Today, just three of these are open to the public for swimming in the summertime. Almost all of them are sited close to the shoreline. Short rail and cable connections ferried stone between the quarries, stone yards, and docks. Whereas competing mainland quarries had to invest in extensive rail systems to transport blocks to the water, Vinalhaven and other Fox Island quarries offered an almost direct connection between quarries and schooners. This coastal advantage helped Maine to lead the granite industry in New England, by the early twentieth century producing more granite products than any other state besides Wisconsin. For every seven granite paving blocks produced in the United States, one came from Maine.[11]

Figure 2.8 **Indian Creek Bridge, Vinalhaven, Maine, 1896**
Source: Courtesy of Vinalhaven Historical Society.

Large quarries, such as Sands, East Boston, and Armbrust Hill, produced a range of products from carved stone to paving blocks. A couple of the other major quarries, like Pequoit and Duschane Hill, specialized entirely in paving blocks, and others changed their product focus over time.[12] More difficult to identify are the countless, nameless "motion quarries" distributed throughout the island. Rather than the large quarries owned by a corporation and operated by a large team of workers, they were often owned by Vinalhaven residents and operated by individuals or teams of two who would cut and sell the blocks to large granite companies by the thousand. Rather than centralized and massive, motions were dispersed and small; rather than deep, they were shallow. Rather than removing large blocks by derrick to be further broken down, blocks were cut directly from the ground. The etymology of "motion" isn't definitive. Dale, however, speculates that motion quarries were named for their "simple and temporary" character.[13] Today, these motions are invisible in the landscape – infilled and overgrown.

We see a range of curious landforms, and sometimes it is not clear whether they resulted from geological processes or human industry. Like eskers, drumlins, and glacial till deposits were formed by migrating Pleistocene ice sheets, these granite landforms are the result of the people quarrying rock. There are piles of rock too large to climb. There are beaches of grout – broken, leftover waste material – that have been colonized and camouflaged by vegetation. There are cairn-like piles of paving blocks that didn't get sold because they were imperfect or not quite squared. There are retaining walls artfully constructed from all classes of waste rock. Part entropy, part craftsmanship, the island's very substance has, over time, been reorganized, dug out, dumped, re-stacked, and re-settled.

From a peak just north of the town center, we peer into Sands Quarry, one of the island's most prolific and impressive. We stand in the backyard of a resident whose house rests on its brink. The ledge that we stand on is the southeast

Figure 2.9 A "motion" paving block quarry

Source: Photograph by Merrithew (T. N. Dale, *The Commercial Granites of New England*. Geological Survey Bulletin; 738. Washington, DC: Government Print Office, 1923, Plate XVIII-B).

Figure 2.10 Grout landscape, Vinalhaven, 2013

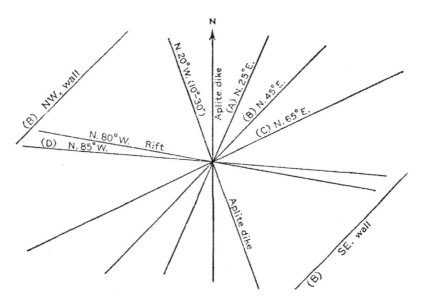

Figure 2.11 **Quarry structure diagram, Sands Quarry, Vinalhaven**

Source: Dale, The Commercial Granites of New England, Plate XVIII-B.

wall, the joint of the quarry. In Dale's *Granites of Maine*, a diagram of Sands Quarry shows how the quarry's primary joints are almost perpendicular to its "rift," the oriented microfoliations that make for clean splitting and determine the overall direction of quarry operations.[14] All bare faces of the quarry align with the joints. In parallel with this angle, sheets of granite between 2 and 10 feet deep were split from the edge of the growing quarry. A U-shaped chasm below is filled with water, strewn off-cuts and grout settling at the bottom. It is illegal to swim here in the summer, but it is idyllic: sheer cliffs, crystal-clear water, and a fascinating cross-section of all that has been removed.

Quarries are self-fulfilling prophecies. When siting a quarry, there is no quarry; yet the more quarry activity, the more quarry-like it becomes. At Sands Quarry, the surface was first marked by a derrick, a rotating crane tethered to the rock; its movable arm and lifts were first powered by oxen and men walking away from its center, and then later by engines. Large pieces of stone could be rotated and placed on a cart or railcar, then transferred to the stone yard for further dissection. Like a compass drawing circles in the rock, the derrick sets in motion the extraction of stone. By the time Dale described the Sands Quarry forty-five years after it opened, an entire hillock had been removed.[15] A hole about 500 by 500 feet, with an average depth of 40 feet, had been excavated. There were five derricks run by three engines, two surface traveling cranes, two steam compressors, four pneumatic drills, a granite lathe capable of working 30-foot columns, and a large stone yard and shed for carving that extended the operations towards the harbor.[16]

I imagine all of the paving blocks, lintels, dressed granite, columns, and carved eagles, wrested from streets and institutions in Washington, DC,

THIS VERY EARLY PHOTO SHOWS THE SANDS QUARRY AND YARD BEFORE
THE BIG SHED WAS ERECTED AND MANY BUILDINGS WERE BUILT.   C 1875

Figure 2.12 **Sands Quarry, Vinalhaven, 1875**

Source: Courtesy of Vinalhaven Historical Society.

Figure 2.13 **Sands Quarry, Vinalhaven, 1907**

Source: Photograph by T.N. Dale (U.S. Geological Survey Bulletin 313: 1907. U.S. Geological Survey Department of the Interior/USGS).

Figure 2.14 **Sands Quarry, Vinalhaven, 2013**

Baltimore, Buffalo, Boston, and New York City, magnetically drawn back to their original position in the granite hill, like a film played backwards. The reunion would be a reverse history of nineteenth-century construction in the northeastern United States. The water now filling the deepest cavities of the quarry would be replaced by blocks removed just before the quarry closed in 1922. Halfway up would fit the pieces cut in the late 1870s. And all the way at the top would fit the blocks extracted in the 1860s, extending the surface that Bill and I stand on as a solid granite hill. Throughout the reassembled hole, millions and millions of hand-cut paving blocks would interlock, their sharp edges melding back together. Packed in with the masses of grout that surround the quarry, all of these objects would create a three-dimensional sculpture, the work of hundreds of quarryman, granite cutters, paving cutters, artisans, including the un-recorded work of women.

To Move an Island by Hand

Stone splits with the greatest ease first along the rift, and then along the grain that runs perpendicular to it. Like a quarryman, a paving cutter reads the rift and grain of the stone in order to determine the path of least resistance to take it apart. The rift may be invisible to a layperson, but to a skilled paving cutter

Figure 2.14 (**Continued**)

it is the map of action. From the quarry, large, rough blocks of stone would be delivered by rail to paving cutters at a stone yard nearby. To turn the bedrock on which you stand into identical units perfectly weighted for one-handed installation, a paving cutter first drills holes along the rift. Large slabs are split off along the rift, their faces becoming the top or bottoms of future paving blocks. Next, the cutter splits along the grain, perpendicular to the rift, producing a surface which becomes the small ends of paving blocks. Subdivision occurs until a slab just one paving block thick remains. This one is separated into individuals with the use of a chisel to mark their intervals, and then split with a whack on the backside.[17] These rougher faces are evened out with a flat-sided hammer. While compressed-air hand drills introduced in the 1890s changed the sound of the stone yard, the methods of paving and stone cutting otherwise changed very little over several decades. However, block dimensions fluctuated constantly. Specifications for New York City paving block contracts changed with evolving tastes, vehicles, and studies on durability. The "improved granite-block pavement" of the 1890s was in the range of 12 inches long, 4 inches wide, and 7 inches deep – proportions optimized for the greatest stability, horse foothold, traction, and consistency.[18]

In one day of splitting, cutting, and trimming, an experienced paving cutter might make 200 nearly identical granite paving blocks.[19] A week's work, or 1,000 such blocks, sold to the quarry owner, would earn between $22 and $30

Figure 2.15 **Paving cutter, Vinalhaven**

Source: Courtesy of Vinalhaven Historical Society.

Figure 2.16 **Paving blocks at Leopold and Company shipping dock, Vinalhaven**

Source: Courtesy of Vinalhaven Historical Society.

(in 1889), depending on if the cutter owned tools.[20] Once sold, a load of blocks bearing the signature of a specific paving cutter's hand was merged into the anonymous flow. Blocks in waiting were stacked into walls or piles: ephemeral fortifications sketching out different architectures to be. Carts on rails would head to the docks, where derricks would lift and place or dump blocks directly into boats. Such rough treatment, one journalist commented, ensured that any inferior granite would be discovered right away; the shipping itself was an abrasion test for the goods.[21] Schooners heading to New York City steered for docks on the East River and the Hudson, where longshoremen offloaded the blocks by hand. Cart boys would then drive the blocks through the streets to sites slated for pavement and deposit them in ubiquitous piles.

Paving blocks were set into the street in a steady procession. First, on top of a hydraulic cement concrete foundation, workers spread clean sand to a depth of one and a half inches thick. No more than 50 feet behind, a line of block setters aligned blocks perpendicular to the street direction. At least 6 feet behind them, a line of rammermen struck the blocks with wood or iron to level the surface, with less than a half-inch margin. Twenty-five feet further, a line of joint fillers then sealed the gaps with gravel, bituminous cement, and Portland cement grout depending on the block's location.[22] Behind the joint fillers, the smooth surface of the street marched down its length. A single paving block set into the street for the long haul, seemingly anonymous with a sea of others, might actually have been in familiar company. New York City specifications called for blocks from similar quarries to be laid together, assuming they would share the same color and grain: "If the blocks are obtained from different quarries or from different parts of the same quarry where the equality or appearance of the rock differs the product of each must be kept separate and laid together on the street."[23] By default, these specifications generated a pixelated map of New England's granite regions – differentiated by variations in color, quality, and workmanship – onto the surface of the streets.

The Economy of Smoothness

This procession of granite in the streets was part of a concerted municipal effort to lubricate the flow of capital. Paradoxically, mid-nineteenth-century New York City was stymied by its own desire to circulate.[24] Efficient transactions and the movement of capital and goods required unhindered circulation, yet all of the means for movement were crumbling under the growth they had promised. If railroads were the arteries of trade, roads were its veins, and unfettered capitalist growth was limited by the roughness of the streets.[25] Stagnation was possible or probable at every single step in the circulation of goods. Dilapidated river docks were glutted with merchandise that needed to move. Streets were jammed with informal markets, narrow and sporadic sidewalks prevented passage, and the piecemeal addition of gas, steam, and other utilities all exacerbated traffic and impeded movement. Horse carriages, automobiles, cable cars, and pedestrians battled for the right of way. Improper

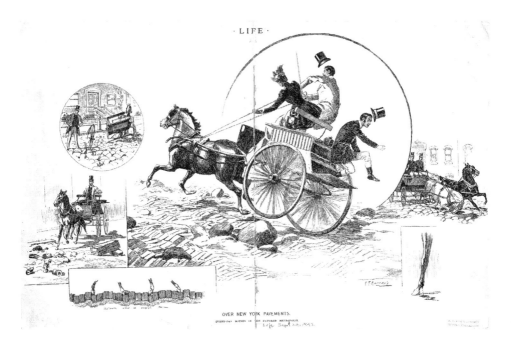

Figure 2.17 "Over New York Pavements: Every-Day Scenes in This Favored Metropolis," 1892
Source: Illustration by Frederick Thompson Richards (Picture Collection, The New York Public Library, Astor, Lenox and Tilden Foundations).

paving festered; stagnant water, mud, and refuse caught in uneven pavements were known to harbor disease and were a menace for new sanitary agendas. Decades of experiments with different materials and installation assemblies left an uneven patchwork of wood, macadam, and stone that sprained horses' legs and sent wagon-riders flying. As David M. Scobey writes: "Lower Manhattan, in short, was a paradox: a perpetual motion machine that perpetually threatened to fall motionless."[26]

In *The Urbanization of Capital*, David Harvey explains how the physical conditions of urban space are related to circulation. As goods and money change hands across space, the speed at which they do so determines the rate of accumulation. The speed at which commodities move becomes as critical as their very cost, and so arrangements to make things move more quickly are essential.[27] Harvey quotes Marx:

> The more production comes to rest on exchange value, hence on exchange, the more important do the physical conditions of exchange – the means of communication and transport – become for the costs of circulation. . . . Capital by its nature drives beyond every spatial barrier. Thus the creation of the physical conditions of exchange . . . become an extraordinary necessity for it.[28]

The notion of "circulation," used to describe the flow of materials and money – like the concept of metabolism taken up by Justus von Leibig and Marx, discussed in the previous chapter – stemmed from early understandings of the human body. Erik Swyngedouw traces the history of the term "circulation" from the influential study of the vascular system in the seventeenth century, through to the metaphors and material conditions of growing metropolises.[29] By the middle of the eighteenth century, money and wealth were said to "circulate" fluidly, likened to the natural systems (the flow of sap, for example) whose scientific study had generated these very concepts. Through the nineteenth century, the notion of circulation was being used to describe urban conditions, from the movement of water for sanitary systems to the speed of traffic to the general organization of space, as Swyngedouw summarizes: "the brisker the flow, the greater the wealth, the health and hygiene of the city would be."[30]

Brisk flow is determined by factors of different scales, from the connectivity of a regional transportation system to the dimensions and properties of a paving block. In a subsection of his 1903 *A Treatise on Highway Construction*, titled "Economy of Smoothness," Austin T. Byrne praises the merits of smooth pavement: "The advantages of smooth pavements to owners and users of horses and vehicles are enormous." Granite was clearly advantageous in this regard. At the same speed and distance, it could take up to eight horses on a poor quality crushed stone macadam to haul the same load that just one and a half to three horses could on granite. With the installation of granite pavement, Bryne argued, "[t]here would be no stuck teams, fewer worried, beaten horses, fewer angry, overworked drivers, and thus fewer delays and interruptions to business."[31] Smoothness felt good, too, and increased the desirability of a city; a visitor "is pleased or displeased exactly in proportion to the smoothness of his journey or the ruggedness of his way."[32] A continuous, smooth pavement was essential for a functional, civilized urban realm, and granite block on an impervious foundation was the "most enduring and economical" material for heavily trafficked roads.[33]

To smoothen New York City's bumpy streets, New York State allotted $5 million over five years towards major repaving efforts, starting in 1888.[34] Thoroughfares were selected for repaving based on an extensive survey of conditions and necessity for public use. Specific material assemblies were assigned to each street segment based on expected traffic: granite blocks on concrete foundation for heavy business traffic, and granite blocks on a sand base for minor business traffic. Asphalt, much quieter under hoof and wheel than granite, was laid in certain residential areas or where noise was perceived to be a problem. For example, asphalt was laid on Broad Street between Wall Street and Exchange Place, and Wall Street between Nassau and Hanover, so that the stockbrokers, lawyers, and bankers might "save their nerves from the strain of the noise of commercial traffic."[35] Plans for new granite surfacing extended throughout lower Manhattan's streets and avenues. Maine was New England's largest producer of granite paving blocks at the time, producing more than 17.5 million blocks in 1889 alone, and therefore supplied a great number of these contracts.[36]

Of the streets slated for repaving, Broadway, New York's famous and infamous thoroughfare, was particularly uneven and jammed up. A *New York Times* correspondent had documented the cacophony in 1882:

> Wagons, carriages, carts, and vehicles of every other description, laden with every variety of burden, throng the roadway; the sidewalks are crowded; the business houses, the offices, and the shops are scenes of the greatest activity; clerks and messengers rush back and forth; grave-looking businessmen hurry from office to office; peddlers and canvassers are making their rounds; excited men and women rush frantically across the street, risking life and limb in their eagerness to gain the opposite side; porters struggle with huge cases of goods being unloaded before the stores; teamsters shout and swear; hawkers cry their wares, and above all rises the rumble, the rattle, and the roar of the ceaseless traffic – in a word, it is Broadway down town at noon.[37]

Not only was the street physically jammed up, the reporter noted, but people of different ethnicities and classes unusually intermingled there:

> [W]omen who are respectable and some who are not; clergymen, speculators, and gamblers; . . . shoplifters and beggars, Frenchmen and Germans, Turks and Russians, Italians and Greeks, Spaniards, Poles, and Chinese. Stretching far along Broadway, this *bizarre* mixture of cosmopolitan life, in which virtue and vice, wealth and poverty, youth and age, the grave and the gay, are oddly intermingled, spreads its panorama.[38]

Under all of this life, Broadway's paving was a mess, as one *New York Tribune* reporter put it: "Ditches and trenches are not novelties here. Sometimes the citizens of New York scarcely know the big business thoroughfare from a hole in the ground."[39]

The Broadway paving contract was enormous. Five miles of granite paving were planned, running from Bowling Green in the south, shooting up north to Union Square, and then darting northwest to the southwest corner of Central Park at 59th Street.[40] In addition to resurfacing with granite paving blocks, a new cable car line would run along its length. After several months of delays and false starts, Broadway's viscera of water and gas mains and cable lines were settled below the street level, ready to be topped with a granite block surface. Over seven months through November 1891, 2.5 million granite blocks were laid down Broadway's length, "the largest number of granite blocks ever laid in the same space of time on one contract,"[41] according to one reporter. The project's contractor had purchased all of the granite block from the New York and Maine Granite Paving Block Company. Such large, profitable paving contracts in New York required strong links to the quarries in Maine to ensure a large supply of paving blocks at the right time. The New York and Maine Granite Paving Block Company was established in 1882 to do just that. From his office on Temple Court in New York City, Maine-born businessman John

Figure 2.18 Looking up Broadway from the corner of Broome Street, circa 1860

Source: Robert N. Dennis Collection of Stereoscopic Views, Miriam and Ira D. Wallach Division of Art, Prints, and Photographs, The New York Public Library, Astor, Lenox and Tilden Foundations.

Figure 2.19 Granite paving and construction of cable car line on Broadway, 1891

Source: Photograph by C.C. Langill and William Gray (Photography Collection, Miriam and Ira D. Wallach Division of Art, Prints and Photographs, The New York Public Library, Astor, Lenox and Tilden Foundations).

Peirce orchestrated the transport of granite from his own quarries in Mount Waldo, Maine, and other quarries along the eastern seaboard, including several on Vinalhaven.[42] He operated as both a receiving agent and middleman for other quarries. He also purchased interests in large granite companies in Vinalhaven such as Bodwell Granite, proprietors of the Sands Quarry.[43] The scale of the company's operation allowed Peirce to control a range of Maine quarries, and win large urban contracts. For Broadway, ten quarries, either owned or commissioned by Peirce, were enlisted to supply the quantity of granite blocks needed.[44] By dominating the granite paving block supply for New York City, Peirce became known as the "Granite King."

No work site is as visible, and therefore prone to public criticism, as street paving. Broadway's construction was no different. Delays in construction drew widespread public frustration. Some argued that the paving delays represented a corrupt collusion between the contractor and the Commissioner of Public Works. According to this scenario, the Commissioner would turn a blind eye to delays, and then late in the season, publicly admonish the contractor to hire thousands of last-minute workers close to election time, bringing favor to the Commissioner's Government.[45] In addition to alleged scandals, general complications with sourcing such a large quantity of paving blocks and working with a new cable car system exacerbated the delay. Furthermore, labor disputes began to arise amongst the paving trades in 1891. Paving layers and other granite workers on the Broadway project striked in early August in support of locked-out quarry workers in Maine – a brief episode of labor solidarity that would mushroom the following year.[46] When the first phase of Broadway's repaving was complete in the fall of 1891, it was met with a range of criticisms about the slow installation, poor project management, and the uneven workmanship of the surface.[47] While much of the Broadway paving was completed by late 1891, the remaining stone would be cut over the winter in Maine and installed the following spring.

Co-urbanization of Vinalhaven

While New York City urbanized with the addition of Vinalhaven's granite, Vinalhaven urbanized through its subtraction. With the arrival of new contracts, stone-cutters came from Scotland, England, Canada, and later Sweden and Finland, more than doubling the population from 1850 to 1880 to more than 2,800.[48] Once-precarious work became more stable, allowing the transitory population to settle. Paving cutters, primarily single men, had been highly mobile – often moving from quarry to quarry as contracts opened up – but as the industry grew, settling became possible. Paving cutters and granite cutters, shielded from the harsh winter conditions, earned about twice as much as quarrymen.[49] And as the community expanded and diversified, associated industries multiplied. Vinalhaven women became known for the production of seines, fishing nets attached to horses for catching shallow-water fish. Though everyone was somehow related to the granite industry, there was little granite to be found in the lives of Vinalhaven residents. No granite streets and few granite buildings were constructed; today, one commemorative carved granite eagle and one granite horse trough can be seen downtown. They are just specks of granite in an island community built of wood, asphalt, glass, and other materials brought there by ferry.

Lucrative federal granite contracts of the mid-nineteenth century guaranteed a 15-percent profit on costs, incentivizing quarry owners to pay quarry workers more to maximize their profits. When the 15-percent contract was banned in 1877, and companies had to submit competitive bids instead, they sought profit through the laborers themselves.[50] Profit could be gained in different ways that pervaded the lived reality of workers and their families. First, labor agreements increasingly disadvantaged workers, enforcing irregular payment, seasonal precarity, and long hours. Second, new mechanized equipment increased production but filled work areas with stone dust, which filled workers' lungs, leading to silicosis. As pneumatic tools gained popularity, death rates from lung diseases followed; by 1918, granite cutters were ten times more likely to die from pulmonary tuberculosis than the general population.[51] Third, quarry companies institutionalized their presence in workers' daily lives through company stores, sometimes paying workers in store credit instead of cash and charging inflated prices for necessities.[52]

Seeing the disparity between quarry owners' profits and their own struggles to survive, workers in Vinalhaven and surrounding areas began organizing,

Figure 2.20 **Bodwell Granite Co. cutters marking the government contract (1872–1888) for the State, War, and Navy Building in Washington, DC**

Source: Courtesy of Vinalhaven Historical Society.

Figure 2.21 **Paving Cutters' Union, Branch No. 34, Vinalhaven**

Source: Courtesy of Vinalhaven Historical Society.

meeting first on nearby Clark Island in 1877.[53] The first Granite Cutters' Union was established, followed shortly after by a Paving Cutters' Union. Within a decade, branches had been formed in almost every state, and membership grew to nearly 6,000.[54] Continual attempts by quarry owners to harass, lock-out, and replace union workers with non-union labor were common but unsuccessful throughout the northeastern quarries. By 1878, Bodwell ordered the dismissal of thirty prominent union members on its payroll, provoking a walk-out and eventual strike.[55]

As quarry workers organized, quarry owners organized to counteract. John Peirce of the New York and Maine Granite Paving Block Company led a consolidation of quarry owners, the Granite Manufacturers' Association, to undermine the granite unions throughout New England. Peirce, Booth, Bodwell, and other heavyweights met to set the price of granite, as well as contract conditions for their workers. Decisions were increasingly made at the Association level, rather than at the quarry level. This meant that the Association could control quarries more likely to compromise with unions, and present an industry-wide tough line on labor. Unionized paving cutters viewed this as an aggressive attempt to minimize wages and wreck the unions by keeping workers precarious.[56]

The Great Strike of 1892

Each year, the Granite Manufacturers' Association of New England would release a set "bill of prices" or annual contract, to be signed by the Paving Cutters', Granite Cutters', and Quarrymen's unions on May 1. But in the spring of 1892, the Association declared that bills of prices would go into effect on January 1, rather than May 1 as before. Laborers would be forced to accept disadvantageous contracts in the winter when work was sparse and they were most vulnerable. The Paving Cutters refused to sign the bills, arguing that the plan was intended to make work more precarious and therefore dismantle the unions.[57] The Manufacturers' Association ordered the quarries to lock the workers out. By May 2, the majority of the 2,000 paving cutters in New England were locked out of twenty-two quarries, with Penobscot Bay being the most affected.[58] Soon after, the Association informed the other granite-related unions, including the Granite Cutters and Quarrymen, that the same contract shift and threat of lockout also applied to them; they too refused to sign, were locked out, and went on strike.

By May 14, some 25,000 quarry workers in total – including paving cutters, granite cutters, quarrymen, as well as blacksmiths and others – were locked out or walked off quarry sites in solidarity.[59] One thousand of these men were from Vinalhaven, the others from quarries throughout Maine, Massachusetts, Rhode Island, Vermont, Connecticut, and New Hampshire.[60] Even with this widespread quarry strike, unions were aware that quarry owners had enough material stockpiled to fulfill their paving and building contracts for some time. The only way for the strike to disrupt these contracts was to stop the flow of granite blocks themselves – from quarry to street and every step in between. Granite unions entreated the unions of all laborers who handled the unloading, transporting, and laying of paving blocks, extending to distant urban centers. A similar action had taken place the year before due to wage disputes, interrupting, among other projects, the repaving of Broadway.[61] Because the New York City region was the major destination for granite paving blocks, the Paving Cutters' Union established its headquarters there in Union Square, forming a "general council of the stone industry" with other granite-related trade unions, including paving layers, rammermen, and stone handlers.[62]

Hundreds of miles from the quarries, on the morning of May 16, 1892, the flow of granite into New York City Streets abruptly stopped. "Pavers, Rammermen, and Stone Cutters Quit Work. – Street Paving Practically Stopped Here and In Brooklyn for the Benefit of Maine Quarrymen – Strikers Confident of Success," read a *New York Times* headline.[63] At the 96th Street East River dock in New York City, schooners laden with Maine granite from Peirce's quarries lay idle as the Longshoremen Union workers refused to unload the boats.[64] Carts, the typical relay method from boats to construction sites, were static. The Cartmen's Union blocked the delivery of locked-out granite to building construction sites.[65] Piles of boycotted granite paving blocks, punctuating Manhattan and Brooklyn, stood deserted, except for a lone watchman. Broadway, Bowling Green, Park Place, Amsterdam Ave., Third Ave., Eleventh Ave.,[66] East and West 138th, West 158th, and West 145th Street lay ripped up and exposed as viscous mud routes, trapping wagons and trucks trying to make their way through.[67] Paving layers, stone cutters, and rammermen, represented by District Assembly No. 49 Knights of Labor in New York City, refused to handle or install the boycotted paving blocks in the city streets.[68] A union circular from the previous February had clearly stated the position:

> Dear Sir. – Desiring to promote harmonious relations between ourselves and our employers, to obviate trouble and prevent future strikes, we, the representatives of the New York, Brooklyn and vicinity Pavers, Rammermen, Stone Handlers' Unions, and Paving Cutters' International Union, in conference assembled, desire to duly notify you, believing it to be fair and just to you and to ourselves to do so, that we do not intend to handle non-union product of our industries in the future.[69]

Simply put, all those tasked with moving granite refused to.

The geographic source and labor practices associated with a pile of paving blocks are anonymous to the passing eye. No labels or identifying trait marks union versus non-union cut stone. At the time of the Great Strike, however, these factors determined whether paving stones were laid or not. Not only were distant labor injustices indicated by the piles of inactive paving stones, they were spatialized throughout the city streets; the more streets slated to be paved with the non-union stone, the greater a disruption to the flow of the city. A resident, traversing across the city, would have a sense of the scale of the resistance. While paving blocks might be the unnoticed underlay of the city, at this time they and their very production were manifest in the streets of New York. Only at Grand Street and 6th Street, where a paving contractor was using blocks from his own, union-operated quarry, did work continue. Like the broken pavements of past experiments, the streets were roughened – a friction created through solidarity with quarry workers far away.

Beyond New York City, in Albany, Baltimore, Boston, Chicago, Philadelphia, Providence, Troy, and other places, an estimated 50,000 to 100,000 workers associated with the locked-out granite workers and walked off job sites.[70] "Nearly all the principal works in course of construction in the United States will suffer," the *New York Times* reported.[71] As granite was the material of choice for symbolic, institutional, and governmental constructions, it was implicated in all of the most important projects. The construction of important symbolic buildings and monuments came to a halt. Four hundred men working on Albany's new Capitol Building, 800 men working on the National Library in Washington, DC, and numerous others working on Brooklyn's Grant Monument, Central Park's Metropolitan Museum of Art addition, and Prospect Park's Memorial Arch left job sites and piles of non-union granite blocks.[72]

The torn-up streets also stymied business. Customers couldn't or wouldn't navigate through mud and piles of paving blocks, and the delivery and transport of other goods was completely obstructed. Business owners petitioned the city and threatened to close until the strike was over.[73] While Commissioner Gilroy ordered paving contractors to fulfill their contracts, even if contractors could locate acceptably sourced granite blocks, there was no workforce that would lay them.[74] Contractors claimed they had no ability to end the strike – it could only be resolved by the New England quarry owners.[75] Peirce claimed that he could easily supplant union labor with non-union labor, namely German workers. However, the league of German pavers promptly rebutted: "This is not a struggle of nationalities, but a struggle to resist the oppressive attempts of the Contractors' Association [the Granite Mfrs. Assoc]."[76]

"There is no strike to speak of in this city," John Peirce told the *New York Times* days after the strike began, even stating that men were working "everywhere," which the journalist refuted. "What trouble there is is at the quarries hundreds of miles away from here,"[77] Peirce told the paper. While Peirce himself was the single-most strategic link between Maine quarries and New York City streets, he understood the ideological power of conceptually separating the two. He sought to distance the construction and modernization of New York City from the labor disputes occurring in Maine to influence the public imagination. New York paving layers stayed on strike until July 11, when they negotiated with paving contractors – now pressed by the city to lay old, used pavers back in place, to rehire union workers, and agree to union rules.[78]

While the workers in New York City went back to work, the New England granite workers continued their strike. "Right here in New England is going on an industrial war besides which that at Homestead becomes rather a small affair,"[79] a *Boston Globe* reporter wrote. In comparison to Homestead, the dispute between the Amalgamated Association of Iron and Steel Workers and the Carnegie Steel Company that had occurred the same summer, the granite strike was generally peaceful and had drawn fewer scabs and hence little violent conflict. By August, some quarries were known to be nearing bankruptcy, and the workers had lost $5 million in income and $300,000 in union expenses.[80] Although the strike had been successful in disrupting the street paving work, coalescing the labor movement, and bringing awareness to their struggle,[81] Paving Cutters' Union secretary James Grant later reflected:

> These methods proved useless; for, although the work was stopped for a time, this stoppage threw all the pavers and rammermen in the city idle. . . . The folly of these things became apparent, and, instead of continuing to waste energy and suffering hardships by idleness, the men turned to and began work for themselves.[82]

Rather than idling, the Paving Cutters' Union pursued co-operative business models in which cutters could continue to supply high-quality blocks to urban contractors. Granite paving, unlike other industries, had advantages for such a plan; granite is ubiquitous, and little capital is required beyond the skill and expertise of the cutters and their tools. Skilled cutters knew how to identify excellent quarries and cut high-quality blocks equivalent to large-quarry stock. "As each man receives the full market value of his labor," Grant wrote, "the men are enabled to make twenty per cent more under the co-operative plan than under the other systems."[83] With the union support of the associated paving layers, rammermen, and longshoremen, the co-operative granite companies could supply high-quality blocks to the city at similar cost – and without exploitation. The city wrongly believed, Grant argued, that only a few large manufacturers were capable of supplying quality paving block. Brooklyn gave the co-operatives their first official contract. From September 1892 to the following January, co-operative quarries on Clark's Island, Tenants Harbor, and other locations provided 1.25 million blocks to the Department of City Works.[84] On Vinalhaven, several co-operative granite companies were formed.[85] By then there were thirteen co-operative quarries running with more than 500 workers employed, and they had supplied 2.5 million paving blocks throughout New York and other cities.[86] They gained credibility and could hardly keep up with contracts. Compared with the Quarrymen who eventually settled with quarry owners, the Paving Cutters claimed victory based on their co-operative ventures. "Beyond all pecuniary profit, the co-operative movement will, in our trade, prevent strikes and lockouts in the future," James Grant proclaimed, "and never again will we see such conditions as those which prevailed in 1892."[87]

The quarry lockout lasted for seven months and was unceremoniously resolved in December of the same year. As one *New York Tribune* reporter wrote:

> It ended much as it began with a few trifling concessions on each side, but for many months it hampered contracts, caused vast annoyance to the building trades and even blocked the triumphant procession of John D. Crimmin's army of workmen in the re-laying of the Broadway pavement.[88]

The Great Strike of 1892 was less a fight about wage or hours and more about checking the power of the Granite Manufacturers' Association, which wielded undue control over granite workers from the quarry to the streets. Union workers shut down the entire material stream where it was quarried, cut, shipped, carted, installed, and finished. By drawing all granite operations to a halt, an otherwise invisible and anonymous material flow became seen and named.

When workers in New York City walked off job sites in support of quarry workers in Maine, the press, the public, and significantly, the Granite Manufacturers' Association, were surprised. While physically distant, New York City unions recognized the Maine granite workers' struggles as contiguous with their own. Collectively the unions displayed that they held significant power over construction operations, which extended geographically throughout the region. The development of co-operative paving quarries empowered workers to rethink the material flow as one not simply to be stopped, but fundamentally reorganized. By the idle piles of paving blocks throughout the city, the unions also communicated labor issues to the public in a physical manner. The strike reminded onlookers, stuck in traffic or walking through piles of inactive blocks, of the thousands of hands, always invisible by distance, involved in the making of the city. Granite, promising smoother streets and capital flow, was an especially strategic flow to stop. Streets are the typical site of public demonstration and social resistance to power, and in the Great Strike of 1892, granite streets were not only the site but also the subject of the conflict.

Peak Granite

Two years after the Great Strike, to protect New York cutters from competition with "cheap labor" from Maine, the New York legislation passed the Tobin Law, which determined that all stone used in state and municipal projects had to be produced within the state. When the New York City Board of Public Works ruled that this law included paving blocks, Bodwell fired all seventy-five of its Vinalhaven cutters.[89] The Paving Unions of New York unanimously opposed the law, claiming that it was a political move to encourage the adoption of asphalt in lieu of granite.[90] Later the same year, when the law was amended due to widespread opposition, Vinalhaven's industry entered a new boom time. Fluctuating demand for masonry construction over the next twenty years produced boom–bust conditions and precarious employment for the granite cutters of the Fox Islands. Such fluctuations not only changed the number of jobs available, but they also altered the ways in which companies offloaded this precarity onto workers. The granite paving industry in Vinalhaven trailed off by the 1930s, with the final large paving company, J. Leopold and Company, closing operations in 1939.[91] Other granite production also decreased substantially in the new century. An official decision by the Treasury Department in 1917 to use Indiana Limestone for new federal buildings halted the widespread specification of granite for big government contracts.[92] Alongside limestone, "artificial stone" (concrete reinforced with steel) also gained prominence in new government projects. In addition, as rail connectivity increased throughout the century, coastal Maine's water access became less of a competitive advantage. Quarrying continued on the islands sporadically until the end of World War II, leaving behind a landscape of abandoned quarries.

During the last decade of the nineteenth century, paving contract specifications revealed the material sea change occurring. Earlier in the decade, the majority of street paving projects were slated for granite with fewer planned in asphalt, but by the turn of the century, asphalt had overtaken. Of Manhattan's 400 miles of paved roads, 156 were paved in granite, and 130 were paved in sheet asphalt. In terms of rate, however, asphalt was quickly gaining lead – nearly all of the 130 miles of asphalt had been paved in less than a decade.[93] A journalist reported that granite was quickly being replaced by asphalt, "which in the opinion of the street officials will cover by far the greater part of the streets of the island within a very few years." Even on Broadway, which had been paved in granite only eight years earlier, asphalt was being added from 47th to 59th Street and from 24th to 26th Street.[94]

In an article titled "Street Paving in America," published in 1893, William Fortune celebrated the new sheet pavements: "It is the most completely distinct modern product of experimental paving; its chief advantage over the block system being a surface of unbroken smoothness, which offers little resistance to traction."[95] While granite pavements had for much of the century promised the smoothest road surface, asphalt trumped all. Granite streets were made of multiple blocks composed of one material; sheet asphalt streets were essentially a single surface, poured in place, composed of many materials. The first asphalt laid in New York City was made of sand and powdered carbonate of lime, mixed with Trinidadian asphalt, refined with petroleum oils. This latter material came largely from the 115-acre "Pitch Lake" in southwestern Trinidad where pitch forms an undulating landscape through which shallow water circulates. In 1891 alone, over 64,000 tons of Trinidadian asphalt made its way to American streets.[96] The transition from solid stone blocks to asphalt brought new issues for street construction. Asphalt's viscous, liquid origins were visible on the streets. On hot days, it could show the markings of horse hooves or wheels, and on cold days it could break apart.[97]

As Jeffrey T. Schnapp argues, the modern fantasy of "frictionlessness" manifested with asphalt's eventual adoption. Whereas engineers had long strived to reduce friction between road and wheels, with asphalt they now had to increase traction to prevent slippage.[98] Asphalt was both modern and prehistoric; as Schnapp writes, it was a "techno-primitive compound" contradictorily tied to progress and to deep time, "mobile but fixed, filthy but clean, archaic but of the future."[99] Silent and sensuous, asphalt also brought a new unsettling smell and uncomfortable heat to the feeling of walking on the street. As asphalt replaced granite in the streets of New York City and other urbanizing centers, supply trajectories shifted; the flow of granite from New England tapered, and channels to Trinidad opened up. One Paleozoic material was replaced with another. If granite hardened the capitalist street to produce a grand infrastructure of continuity for the nineteenth century, asphalt softened the street, poured and rolled into place for the twentieth.

One of the last "motion" paving quarries to close on Vinalhaven belonged to Fred Chilles, Bill's great uncle. A photo that Bill took shows an abandoned derrick with an osprey nest perched on top of it, and the forest encroaching in from all sides. Now imperceptible from the road, Chilles' quarry has merged back into the land, unlike the larger ones. Sands Quarry (where Granite King Peirce purchased blocks from), Booth's Quarry, and Lawson's Quarry are now popular swimming spots in Vinalhaven, some legal, others not. The stepped ledges that allowed for the smooth extraction of large hunks of stone are now ideal platforms from which to dive. Today, Vinalhaven is an active fishing community, with a population similar to its pre-granite days. Some of the larger houses are the summer homes of families passed down from the days of big granite.

One small quarry operates on Vinalhaven today, a "hobby" quarry run by four gentlemen who call themselves the "Four Fossils," a poke at their age.[100] Setting up shop in an overgrown quarry that had been closed for forty years, the fossils – a retired accountant, a heavy equipment operator, a real estate agent and sculptor, and a boot company president – were drawn for different reasons to make something of the dormant granite and quarry landscape.[101] An

Figure 2.22 **Fred Chilles' quarry, 1994**

Source: Photograph by Bill Chilles (courtesy of Vinalhaven Historical Society).

estimated 45,000 cubic feet of granite blocks are still stacked on the site, stopped in time from the quarry's closure and the fact that 16,000-pound rock pieces aren't easy to pilfer. Using scavenged, retrofitted industrial equipment, the Fossils custom-cut stone for local landscaping, architectural, and sculptural use. When new construction projects on Vinalhaven incorporate granite quarried outside of the island, the public is not impressed.[102] The irony of purchasing granite from elsewhere is too sharp to ignore, and the Four Fossils, in small quantities, cut Vinalhaven granite for Vinalhaven. The Fossils see their work as a project for enjoyment not profit, and no one speaks of another granite age to come.

Granite flow between Maine and New York can be seen as a series of motions: the flow of slowly cooling stone; quarrymen and paving cutters

Figure 2.23 **Lawson's Quarry, Vinalhaven, 2013**

removing blocks from granite ledges; the transport of granite blocks from island quarries southward; the laying of countless granite blocks in New York City streets; the mobility afforded by consistent stone surfaces; and the movement of the granite industry overseas. As Marina Fischer-Kowalski suggests, the concept of material flow, derived from various scientific fields, has been productively utilized by social theorists, anthropologists, and social geographers to study socio-ecological processes.[103] From ecological economics to industrial ecology, the concept is promising for examining the inextricable relationships between landscape change, capitalism, and labor.[104] In material flow language, materials are the protagonists, the actors, the nouns. While this is enticing, as it animates material otherwise considered inert, it tends to erase the social dimensions of that material flow. The story of granite flow between Vinalhaven and New York City resists the notion of an abstract, anonymous, invisible material flow. Rather than abstract, the substance described here is heavy, textured, and staunchly material. Rather than anonymous, the granite moving here was cut, transported, laid, and finished by thousands of workers who, during the Great Strike of 1892, demonstrated their essential role in this flow by shutting it down. Rather than invisible, the flow of granite from Vinalhaven to New York City's streets is expressed in the depth of the swimming hole that kids dive into today.

Notes

1 Piotr Migoń, *Granite Landscapes of the World* (Oxford: Oxford University Press, 2006), 9.
2 Thomas Nelson Dale, *The Granites of Maine*, United States Geological Survey Bulletin 313 (Washington, DC: Government Printing Office, 1907), 20.
3 Ibid., 25–6.
4 Ibid., 12.
5 Ibid., 67.
6 Carleton A. Chapman, "Bays-of-Maine Igneous Complex," *Geological Society of America Bulletin* 73, no. 7 (1962): 887.
7 Dale, *The Granites of Maine*, 129.
8 George Rich, "The Granite Industry in New England," *New England Magazine*, February 5, 1892, 748.
9 George Perkins Merrill, *Stones for Building and Decoration* (New York: John Wiley & Sons, Inc., 1891), 184.
10 Bill Chilles, personal communication, November 23, 2013.
11 Dale, *The Granites of Maine*, 12.
12 Ibid., 135.
13 Ibid., 72.
14 Ibid., 131.
15 Ibid.
16 Ibid., 132.
17 "The Manufacture of Granite Paving Blocks," *Engineering News* 73, no. 8 (1915): 380.
18 Samuel Whinery, *Specifications for Street Roadway Pavements* (New York: Engineering News Publishing Company, 1907), 38; Austin T. Byrne, *A Treatise on Highway Construction* (New York: John Wiley & Sons, 1903), 104–5; "Of What Use Are the Specifications: Alleged Violation of Contract in Street Paving," *New York Tribune*, November 11, 1889, 3.
19 "The Manufacture of Granite Paving Blocks," 380.
20 Byrne, *A Treatise on Highway Construction*, 32.
21 "The Manufacture of Granite Paving Blocks," 381.
22 Whinery, *Specifications for Street Roadway Pavements*, 38–40.
23 Ibid., 38.
24 David M. Scobey, *Empire City: The Making and Meaning of the New York City Landscape* (Philadelphia: Temple University Press, 2002), 141.
25 Byrne, *A Treatise on Highway Construction*, 104–5.
26 Scobey, *Empire City: The Making and Meaning of the New York City*, 141.
27 David Harvey, *The Urbanization of Capital: Studies in the History and Theory of Capitalist Urbanization* (Baltimore: John Hopkins University Press, 1985), 36.
28 Karl Marx, *Grundisse*, trans. Martin Nicoloaus (Harmondsworth: Penguin Books, 1973), 524.
29 Erik Swyngedouw, "Metabolic Urbanization," in *In the Nature of Cities: Urban Political Ecology and the Politics of Urban Metabolism*, eds. Nikolas C. Heynen et al. (London and New York: Routledge, 2006), 31.
30 Ibid., 32.
31 Byrne, *A Treatise on Highway Construction*, 7.
32 Ibid., 5.
33 Ibid., 102.
34 "For Paving the Streets: Where the New Work Will be Done," *New York Tribune*, February 25, 1891, 10.
35 "Plans for Street Paving: Commissioner Gilroy Submits a Report," *New York Times*, July 26, 1889, 8.
36 Byrne, *A Treatise on Highway Construction*, 30.

37 "A Study of Broadway: The Great Thoroughfare from Dawn to Dawn," *New York Times*, April 29, 1882, 4.

38 Ibid.

39 "Tearing up Broadway," *New York Tribune*, March 27, 1891, 4.

40 Rich, "The Granite Industry in New England," 762; "Extensive Paving Plans: Mr. Gilroy Wants More Authority," *New York Tribune*, March 7, 1890, 4.

41 "Broadway Paved at Last: The Cable Conduit Is Also Practically Finished," *New York Times*, December 19, 1891, 6.

42 Rich, "The Granite Industry in New England," 762; "No Work at the Quarries," *New York Tribune*, May 17, 1892, 1.

43 William Herries, *Brooklyn Daily Eagle Almanac: A Book of Information, General of the World, and Special of New York City and Long Island* (Brooklyn: Brooklyn Daily Eagle, 1891).

44 "Broadway Paved at Last: The Cable Conduit Is Also Practically Finished," 6.

45 "The Broadway Pavement: It Already Needs Repairing," *New York Tribune*, October 26, 1891, 1.

46 "The Pavers' Strike: It Extends, Much to the Annoyance of Many Interests," *New York Times*, August 2, 1891, 3.

47 "That Broadway Pavement: Why it is Already in Poor Condition," *New York Times*, October 28, 1891, 11.

48 Roger L. Grindle, "Bodwell Blue: The Story of Vinalhaven's Granite Industry," *Maine Historical Society Quarterly* 16, no. 2 (1976): 51.

49 Cynthia Burns Martin, "The Bodwell Granite Company Store and the Community of Vinalhaven, Maine, 1859–1919," *Maine History* 46, no. 2 (2012): 155.

50 Grindle, "Bodwell Blue," 54.

51 Frederick Ludwig Hoffman, *The Problem of Dust Phthisis in the Granite-Stone Industry*, United States Bureau of Labor Statistics Bulletin 293 (Washington, DC: Government Printing Office, 1922), 28.

52 Burns Martin, "The Bodwell Granite Company Store," 163.

53 Grindle, "Bodwell Blue," 54.

54 State of Maine, *Third Annual Report of the Bureau of Industrial and Labor Statistics for the State of Maine 1889* (Augusta: Burley & Flynt, 1890), 45.

55 Grindle, "Bodwell Blue," 58–9.

56 "Beginning a Bitter Fight: The Struggle of the Granite Men," *New York Tribune*, May 16, 1892, 1.

57 New York State, *Tenth Annual Report of the Bureau of Statistics of Labor of the State of New York, for the Year 1892: Part 2* (Albany: James B. Lyon, 1893), 296.

58 Ibid., 295; "Silent Stones: Quarrymen Quit Work in New England," *Boston Daily Globe*, May 3, 1892, 1.

59 "Men Are Firm: Granite Workers' Strike Cost $300,000," *Boston Daily Globe*, August 1, 1892, 8.

60 "Silent Stones," 1.

61 "Street Pavers Will Work Again," *New York Tribune*, August 5, 1891, 7.

62 New York State, *Tenth Annual Report*, 294–5.

63 "Twelve Hundred Men Out," *New York Times*, May 10, 1892, 5.

64 "No Work at the Quarries," *New York Tribune*, 1.

65 Ibid.

66 "Twelve Hundred Men Out," 5.

67 "The Great Strike Begun," *New York Times*, May 17, 1892, 1.

68 "Big Fight of Granite Cutters Hundreds of Men Refuse to Handle 'Lock Out' Paving Blocks Edicts of Granite-Cutters Union," *New York Tribune*, May 10, 1892, 1.

69 New York State, *Tenth Annual Report*, 296–7.

70 "Twelve Hundred Men Out," 5; "Men are Firm," 8; "Strike to Follow Lockout: One Hundred Thousand Men May Quit Work," *New York Times*, May 14, 1892, 1.

71 Ibid., 1.

72 Ibid.
73 "The Great Strike Begun," 1.
74 Ibid.
75 "No Work at the Quarries," 1.
76 "Beginning a Bitter Fight," 1.
77 "The Great Strike Begun," 1.
78 New York State, *Tenth Annual Report*, 298.
79 C.F. Willard, "Granite Fight: Homestead Troubles Pale Beside It," *Boston Daily Globe*, September 19, 1892, 10.
80 "Men are Firm," 8.
81 New York State, *Tenth Annual Report*, 299.
82 Ibid., 299–300.
83 Ibid., 301.
84 "Men are Firm," 8.
85 Grindle, "Bodwell Blue," 86.
86 New York State, *Tenth Annual Report*, 299–300.
87 Ibid., 301.
88 "Labor in New-England: No Disturbances and Universal Prosperity," *New York Tribune*, December 18, 1892, 2.
89 Grindle, "Bodwell Blue," 88.
90 "An Imposition on Workingmen: The Law about Dressing Stone for Public Works Not in the Interest of Labor," *New York Tribune*, February 18, 1895, 5.
91 Grindle, "Bodwell Blue," 103.
92 Ibid., 101.
93 "Many Miles of Pavements: Some Facts Relating to the Streets of the Borough of Manhattan," *New York Times*, December 25, 1898, 21.
94 Ibid.
95 William Fortune, "Street-Paving in America," *Century Illustrated Magazine*, October 1893.
96 Ibid.
97 Ibid.
98 Jeffrey T. Schnapp, "Three Pieces of Asphalt," *Grey Room* 11 (2003): 9.
99 Ibid., 16.
100 Karen Roberts Jackson, "Boys with Toys . . . and Vision," *The Working Waterfront Archives*, July 1, 2008, accessed June 23, 2015, www.workingwaterfrontarchives.org/2008/07/01/boys-with-toys-and-vision.
101 Ibid.
102 Harry Gratwick, "The Return of Vinalhaven's Four Fossils," *The Working Waterfront Archives*, September 1, 2009, accessed June 23, 2015, www.workingwaterfrontarchives.org/2009/09/01/the-return-of-vinalhavens-four-fossils.
103 Marina Fischer-Kowalski, "Society's Metabolism: The Intellectual History of Materials Flow Analysis, Part I, 1860–1970," *Journal of Industrial Ecology* 2, no. 1 (1998).
104 An important contribution to industrial ecology, Material Flow Analysis is a quantitative method of measuring flows and stocks of materials given a particular system boundary, be it a city, country, or the planet.

Bibliography

"Beginning a Bitter Fight: The Struggle of the Granite Men." *New York Tribune*, May 16, 1892.
"Big Fight of Granite Cutters Hundreds of Men Refuse to Handle 'Lock Out' Paving Blocks Edicts of Granite-Cutters Union." *New York Tribune*, May 10, 1892.

"Broadway Paved at Last: The Cable Conduit Is Also Practically Finished." *New York Times*, December 19, 1891.

"The Broadway Pavement: It Already Needs Repairing." *New York Tribune*, October 26, 1891.

Byrne, Austin T. *A Treatise on Highway Construction*. New York: John Wiley & Sons 1903.

Chapman, Carleton A. "Bays-of-Maine Igneous Complex." *Geological Society of America Bulletin* 73, no. 7 (1962): 883–8.

Chilles, Bill. Personal communication, November 23, 2013.

Dale, Thomas Nelson. *The Granites of Maine*. USGS Bulletin 313. Washington, DC: Government Printing Office, 1907.

"Extensive Paving Plans: Mr. Gilroy Wants More Authority." *New York* Tribune, March 7, 1890.

Fischer-Kowalski, Marina. "Society's Metabolism: The Intellectual History of Materials Flow Analysis, Part I, 1860–1970." *Journal of Industrial Ecology* 2, no. 1 (1998): 61–78.

"For Paving the Streets: Where the New York Will be Done." *New York Tribune*, February 25, 1891.

Fortune, William. "Street-Paving in America." *Century Illustrated Magazine*, October 1893.

Gratwick, Harry. "The Return of Vinalhaven's Four Fossils." *The Working Waterfront Archives*, September 1, 2009. Accessed June 23, 2015, www.workingwaterfront archives.org/2009/09/01/the-return-of-vinalhavens-four-fossils/.

"The Great Strike Begun." *New York Times*, May 17, 1892.

Grindle, Roger L. "Bodwell Blue: The Story of Vinalhaven's Granite Industry." *Maine Historical Society Quarterly* 16, no. 2 (1976): 51–112.

Harvey, David. *The Urbanization of Capital: Studies in the History and Theory of Capitalist Urbanization*. Baltimore: John Hopkins University Press, 1985.

Herries, William. *Brooklyn Daily Eagle Almanac: A Book of Information, General of the World, and Special of New York City and Long Island*. Brooklyn: Brooklyn Daily Eagle, 1891.

Hoffman, Frederick Ludwig, *The Problem of Dust Phthisis in the Granite-Stone Industry*. United States Bureau of Labor Statistics Bulletin 293. Washington, DC: Government Printing Office, 1922.

"An Imposition on Workingmen: The Law About Dressing Stone for Public Works Not in the Interest of Labor." *New York Tribune*, February 18, 1895.

Jackson, Karen Roberts. "Boys With Toys . . . and Vision." *The Working Waterfront Archives*, July 1, 2008. Accessed June 23, 2015, www.workingwaterfrontarchives. org/2008/07/01/boys-with-toys-and-vision/.

"Labor in New-England: No Disturbances and Universal Prosperity." *New York Tribune*, December 18, 1892.

"The Manufacture of Granite Paving Blocks." *Engineering News* 73, no. 8 (1915): 376–81.

"Many Miles of Pavements: Some Facts Relating to the Streets of the Borough of Manhattan." *New York Times*, December 25, 1898.

Martin, Cynthia Burns. "The Bodwell Granite Company Store and the Community of Vinalhaven, Maine, 1859–1919." *Maine History* 46, no. 2 (2012): 149–68.

Marx, Karl. *Grundisse*. Translated by Martin Nicolaus. Harmondsworth: Penguin Books, 1973.

"Men Are Firm: Granite Workers' Strike Cost $300,000." *Boston Daily Globe*, August 1, 1892.

Merrill, George Perkins. *Stones for Building and Decoration*. New York: John Wiley & Sons, Inc., 1891.

Migoń, Piotr. *Granite Landscapes of the World*. Oxford: Oxford University Press, 2006.

New York State. *Tenth Annual Report of the Bureau of Statistics of Labor of the State of New York, for the Year 1892: Part 2.* Albany: James B. Lyon, 1893.

"No Work at the Quarries." *New York Tribune*, May 17, 1892.

"Of What Use Are the Specifications: Alleged Violation of Contract in Street Paving." *New York Tribune*, November 11, 1889.

"The Pavers' Strike: It Extends, Much to the Annoyance of Many Interests." *New York Times*, August 2, 1891.

"Plans for Street Paving: Commissioner Gilroy Submits a Report." *New York Times*, July 26, 1889.

Rich, George. "The Granite Industry in New England." *New England Magazine*, February 5, 1892, 748.

Schnapp, Jeffrey T. "Three Pieces of Asphalt." *Grey Room* 11 (2003): 5–21.

Scobey, David M. *Empire City: The Making and Meaning of the New York City Landscape.* Philadelphia: Temple University Press, 2002.

Silent Stones: Quarrymen Quit Work in New England." *Boston Daily Globe*, May 3, 1892.

State of Maine. *Third Annual Report of the Bureau of Industrial and Labor Statistics for the State of Maine 1889.* Augusta: Burley & Flynt, 1890.

"Street Pavers Will Work Again." *New York Tribune*, August 5, 1891.

"Strike to Follow Lockout: One Hundred Thousand Men May Quit Work." *New York Times*, May 14, 1892.

"A Study of Broadway: The Great Thoroughfare From Dawn to Dawn." *New York Times*, April 29, 1882.

Swyngedouw, Erik. "Metabolic Urbanization." In *In the Nature of Cities: Urban Political Ecology and the Politics of Urban Metabolism*, edited by Nikolas C. Heynen, Maria Kaika, and Erik Swyngedouw, 21–40. London and New York: Routledge, 2006.

"Tearing up Broadway." *New York Tribune*, March 27, 1891.

"Twelve Hundred Men Out." *New York Times*, May 10, 1892.

Whinery, Samuel. *Specifications for Street Roadway Pavements.* New York: Engineering News Publishing Company, 1907

Willard, C.F. "Granite Fight: Homestead Troubles Pale Beside It." *Boston Daily Globe*, September 19, 1892.

Figure 3.1(A) Detail, view looking east
along the Monongahela River towards
Carrie Furnaces, 1959

Source: Pittsburgh & Lake Erie Railroad
Company Photographs, 1886–1972, University of
Pittsburgh Archives and Special Collections.

Figure 3.1(B) Detail, Riverside Park under construction, view looking south from 82nd Street, 1936

Source: Milstein Division, The New York Public Library.

Chapter 3

Rivers of Steel

Steel from Pittsburgh to Riverside Park, 1937

Cycling north on the Hudson River Greenway towards Riverside Park, I follow one of the spaghetti-like routes that wind together as if fighting for the best view of the river. To my immediate right, vehicles snake up the West Side Highway. Beyond, ant-sized figures walk along the elevated steel structure of the High Line, the public park occupying the decommissioned New York Central Railroad line. Where tourists stroll along obsolete rail structures through converted warehouse buildings to visit the new Whitney Museum, cattle and pigs once traveled to slaughterhouses in live freight cars. Once the space of industry, trains, and meatpacking, this riverfront neighborhood now contains some of the most expensive luxury real estate in the country. To my left, the wide Hudson River appears to rush towards the Atlantic Ocean. But as a tidal estuary, the current reverses twice a day, pulsing salt water up and down the lower river.[1] Piers, the strategic ports formerly connecting industry along the Hudson to Albany and Troy, are now parks, tourist attractions, and private sports clubs. A water taxi skirts the shoreline; kayakers and ducks circle and flock towards their own. Around 59th Street, where Riverside Park South – a segment developed in the 1980s[2] – begins, the highway rises from the ground, forming a riveted steel cathedral that dwarfs the basketball players inside. Here, the bike lane slips underneath the highway and then propels me into Riverside Park, which extends from 72nd Street all the way up to 158th.

It is rare to feel so far away from buildings in Manhattan. Here, facing the New Jersey Palisades across the Hudson, it is almost possible to remember that this place is the edge of a big island. As landscape architect Jens Jenson wrote in 1916,

> The bluffs of the Hudson, as represented in Riverside Park, bold and powerful, show the work of ages from the far distant geological times, and are in themselves a masterpiece. They link us to the great past, back to unknown ages. They express the power and force of the elements; they represent a book of knowledge that no living man can produce.[3]

These impressive bluffs set the major contours of the park site, but human construction has altered its profile countless times. Today, you can think of Riverside Park as a cross-section of three terraces. At the top is Riverside Drive;

Figure 3.2 **Riverside Park promenade, looking north, 2015**

in the middle is the promenade, the park's central spine; and at the water's edge are active recreation facilities, including sports courts, the narrow pedestrian and bike path, and, for much of the park's length, the Henry Hudson Parkway (continuing the West Side Highway northward). The park's levels are expertly intertwined with underpasses and grade separations, steps, and walls. The park is designed to maximize use and movement along four miles of Manhattan's prime riverfront.

In the middle promenade level, it is remarkably quiet. I'm sheltered beneath the activity of Riverside Drive, yet perched above the Parkway with a sweeping view of the river. If I hadn't been looking for them, I wouldn't have noticed the steel gratings that dot the length of the promenade. Peering down through a grate, I can see polished steel rail tracks glinting in the darkness below. A 66-foot-wide rail right of way – formerly owned by the New York Central Railroad Company, now by Amtrak – is hidden beneath the promenade. Every so often, a high-pitched train barrels through the tunnel, vibrating the ground, and sending leaves flying. Seconds later, the sounds of birds, muffled traffic, and river flow return. Ninety years ago the same train tracks were open to the air, laden with steam locomotives and live freight, obstructing access to the River, and further south, tangled with the streets of lower Manhattan. But in 1937, a steel structure was installed over the tracks and then covered in stone, soil, and trees, providing a seamless, terraced landscape for a more genteel park experience.

Figure 3.3 **Riverside Park, tunnel grating, 2015**

What one experiences today as a naturalistic and rugged bluff landscape – the edge of a granitic island – is structured overwhelmingly with steel.

If granite smoothed the ground for better circulation in the late nineteenth century, steel multiplied the city's very carrying capacity in the early twentieth. The stacking logic of granite and masonry was supplanted by the tectonic, connective logic of steel. In this three-dimensional proliferation of the ground, steel not only sprouted as towers, it became bridges and other infrastructure that separated antagonistic urban activities. An x-ray of Riverside Park and the entire west side of Manhattan would reveal an island-sized organism with interminable vertebrae of steel girders and columns and limbs of spurs. It rises from the south and then plunges underground around 30th Street, tunneling through Riverside Park, and then emerges until it becomes the Hudson River Bridge at the northern tip of Manhattan.

This chapter follows structural steel from one riverside to another in the 1930s: from United States Steel Corporation's Carrie Blast Furnaces on the Monongahela River near Pittsburgh to Riverside Park on the Hudson River in Manhattan. Steel making profoundly transformed the Monongahela Valley in Pittsburgh, connecting to a network of ore, coal, and limestone deposits in Minnesota and southern Pennsylvania. This very steel physically transformed Manhattan, reorganizing the modern city. The 1936 covering of the rail tracks in Riverside Park was part of the West Side Improvement, a massive restructuring of Manhattan's west side with steel. Some of this structural steel came from the American Bridge Company, a U.S. Steel subsidiary manufacturer, itself likely supplied by the Homestead Steel Mill, whose iron was supplied by the Carrie Blast Furnaces. At this moment in time, New Deal policies reshaped the public sphere: in Riverside Park this occurred as people went back to work under Federal emergency-relief programs, constructing scores of public works while developing new public modern landscapes; and in Pittsburgh through the empowerment of labor and the development of industrial unionism, which introduced the worker to the bargaining table. This chapter looks at the role of structural steel and progressive legislation in the 1930s in constructing a new public landscape and in supporting organized labor.

Two Riversides

As the only all-rail connection bringing produce and meats into the city, the New York Central and Hudson River Railroad's (later shortened to New York Central RR) West Side Line struck a course through Riverside Park until the mid-1930s.[4] While Riverside Park legally extended beyond the tracks and into

Figure 3.4 **Riverside Park, view looking north towards Grant's Tomb, rail tracks visible, 1901**

Source: Photograph by William Henry Jackson (Library of Congress Prints and Photographs Division).

Figure 3.5 Riverside Park showing industrial uses along the river, view looking north from 72nd Street, 1915

Source: Milstein Division, The New York Public Library.

the water, the Hudson Riverside was physically isolated and therefore easily appropriated by private interests. Coal and material storage, timber berths, garbage piers, and all forms of dumpsites populated the shoreline.[5] Although some of these uses were illegal, the city actively granted permits for private uses on this public land. The tracks created a physical barrier in the park, but they also brought smoke-spewing locomotives and pungent abattoir-bound livestock, whose smells wafted upslope, angering wealthy Riverside Avenue residents.[6] At the time, the formerly working-class neighborhood was attracting wealthier property owners looking for more rustic accommodation. Among the new residents were industrialists amassing fortunes in steel and rail, notably Charles Schwab, who would become the first president of U.S. Steel.[7]

Conflicts between industrial and civic uses had long characterized the Riverside site. The park's original landscape architect, Frederick Law Olmsted, addressed the tensions between the industrial waterfront and civic life as a central organizing principle. Olmsted argued that the land along the river was too steep to access and that buildings (presumably industrial in nature) would obstruct views. He concluded that the mid-slope of the tract had the best views and should be structured as a continuous shelf, generous enough to accommodate large shading trees.[8] At the easternmost edge, Olmsted convinced the Parks Commission to extend the proposed park limits to include Riverside Avenue and bundle the pedestrian and vehicular routes as a grand promenade where trees could shade pedestrians and allow for beautiful vistas from the driver's seat.[9] The design concealed the riverside industrial infrastructure and also championed the modern road and the automotive experience. Olmsted's vision was generally achieved over the coming decades, though in a piecemeal fashion. Perhaps due to this slow development, Riverside Park was perceived as incomplete and especially vulnerable to encroaching private interests along the shoreline.[10]

In the early twentieth century, the New York Central Railroad made plans to expand their operations through Riverside Park, further threatening the park's coherence.[11] Citizen groups, such as the Women's League for the Preservation of Riverside Park, held rallies, published editorials, and networked with city officials to fight the proposed expansion and the general selling off of the park to industry. Mobilizing arguments for children's welfare, public health, and the restorative value of nature, the organization battled the privatization of New York's dwindling public waterfront, calling out in one brochure:

> The city has 400 miles of water-front for commercial purposes, only five miles for Park purposes. Do the citizens of New York wish to give this last five miles to the railroad? . . . Do the citizens *want* a Park, or do they want a freight yard, and the view of a cattle runway south of 72nd Street?[12]

Plans for Riverside Park shifted with political tides, but by July 1929, after forty years of struggle between the city and New York Central, plans were approved to cover the tracks in the park and make a new public esplanade, eliminate railroad grade crossings, and introduce a new express highway.[13] Work began, but within months the stock market crashed, disrupting all major construction agendas.

Three-hundred and seventy miles to the west, the banks of the Monongahela River around Pittsburgh were also congested and their use contested. Pittsburgh's steel industry had emerged around the Point, a 40-degree-angled landmass where the Monongahela and Allegheny Rivers converge and spawn the Ohio, but development spread rapidly along the three rivers' shores by the turn of the twentieth century. Pittsburgh offered strategic geographic advantages to steel makers – it was well-situated to growing midwestern markets, close to coking coal in southwestern Pennsylvania, and had good access to Great Lakes ore routes, as well as barge and rail transport along its three rivers.[14] Riverside sites were constrained, though gracious river meanders provided many deep sites along floodplain deposition banks. Large steel works and blast furnaces quickly colonized these locations, alternating along the river's edge.[15]

As Pittsburgh ascended as an industrial center, steel production transformed the shape of the rivers, as well as the colors of the water and sky. The Monongahela was soon considered the country's "most used river," bearing more cargo than even the Panama or Suez Canals.[16] River waters were canalized and dammed to regulate water levels, river bottoms were dredged to allow the passage of ships, and riverfront parcels extended into the river with landfill. These manipulations – expertly engineered for industrial utility – also destroyed biological communities at all elevations, from alluvial floodplains, to wetland sedges and grasses, to river-bottom fish and shellfish habitats.[17]

The Monongahela River received toxic industrial waste, acid mine drainage, and raw municipal sewage – and yet it served as an unfiltered municipal drinking water source until 1907. At the turn of the twentieth century, typhoid death rates in Pittsburgh (stemming from river contamination) were nearly triple the national average and affecting African American and new immigrant communities the most severely.[18] Thick black smoke from industries along the rivers led many to liken the city's vista to hell. Pittsburgh's high death rate due to air pollution made it, as one study claimed, an "ideal laboratory" for pneumonia research.[19] Working-class residents lived in deplorable conditions. Journalist Henry Mencken characterized Pittsburgh's contradictions in 1927:

> Here was the very heart of industrial America, the center of its most lucrative and characteristic activity, the boast and pride of the richest and grandest nation ever seen on earth – and here was a scene so dreadfully hideous, so intolerably bleak and forlorn that it reduced the whole aspiration of man to a macabre and depressing joke.[20]

Figure 3.6 **Smoky sky in Pittsburgh, looking north to the Allegheny River from the roof of Union Station**

Source: Photograph by Adam M. Brown, American (1888–1970), "Mills in Strip District, Pittsburgh," 1906, gelatin silver print, 8 in. × 10 7/16 (Photography © 2019 Carnegie Museum of Art, Pittsburgh).

As industry staked out the riverfronts, residents had little access to the increasingly toxic and obstructed rivers, and began to perceive them as strictly industrial.[21] Frederick Law Olmsted Jr. (son of Frederick Law Olmsted Sr.), hired by the Pittsburgh Civic Commission in 1910 to develop a city-planning document, argued that the riverfront had immense potential for recreation.[22] He challenged the public to question their beliefs that the functional and industrial activities of the waterfront were ugly and incompatible with civic life.

> It does not diminish the essential grandeur of the situation that the river swarms with barges and steamers; that it is spanned by busy bridges; that the flat lands along the river are crowded with railroads, buildings and smoking factories . . . It is a spacious and impressive landscape in any case.

But nevertheless, according to Olmsted Jr., to appreciate the river, even in its industrial character, the public had to be given physical access to it, and the land had to be treated with respect.[23] Olmsted Jr. mapped out the densifying industrial developments along Pittsburgh's riversides and argued that the city should maintain some riverfront land for public parkland, including Nine Mile Run, a valley not far from the Homestead Steel Works. The political

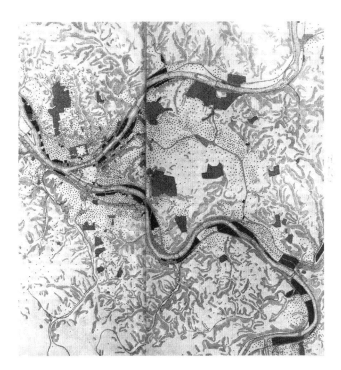

Figure 3.7 **Detail, map showing manufacturing properties (in darkest tone) along the Monongahela River (bottom) and Allegheny River (top) in Pittsburgh**

Source: Frederick Law Olmsted Jr., *Pittsburgh Main Thoroughfares and the Downtown District*. Pittsburgh: Pittsburgh Civic Commission, 1910.

administration sided with business interests and largely ignored Olmsted Jr.'s recommendations, neglecting to acquire new parkland on the rapidly industrializing waterfront.[24]

Territories of Steel

Ron Baraff appears from across the meadow that flanks Carrie Blast Furnaces 6 and 7.[25] We're in Swissvale, seven miles east of downtown Pittsburgh, on a northern bank of the Monongahela River, taking in the towering furnaces. Across the river is the former steel mill site, which together with the Carrie Blast Furnaces comprised the Homestead Steel Works, operated by Carnegie Steel until 1901 and then by U.S. Steel until 1986. For seventy years, a near continual stream of molten iron flowed from Carrie's furnaces, crossed the river by bridge, entered Homestead's mills, and left as structural steel members and plates. Sprawling for 420 acres along both north and south banks of the Monongahela, the Homestead Works exemplified the principles of large-scale integrated industrial works that defined American industry in the twentieth century.

I'm there because I'm trying to figure out where the steel that was used to cover the tracks in Riverside Park in the 1930s came from. It's not information that is particularly important to anyone else, but it is a structural aspect

Figure 3.8 **Carrie Furnaces 6 and 7, looking through the former ore yard, 2014**

of this book, so I'm giving it my all. After poring over newspaper articles, steel orders, and industrial archives, I found a small clue that the American Bridge Company, a U.S. Steel subsidiary structural steel manufacturer, supplied a significant load to the West Side Improvement to cover New York Central tracks on the Hudson at the same time of the Riverside Park renovation, in 1936.[26] It isn't conclusive (I don't know if this steel went directly inside Riverside Park, or was installed further south), but it is the evidence that I decide to work with. Ambridge, the town where the American Bridge Company's headquarters used to be, is not far down the Ohio River from Homestead and Carrie Furnaces, and shipments of steel from Homestead were regularly sent to Ambridge by barge and rail. When I ask Ron if he thought the 1936 Ambridge shipment that I was interested in might have included iron from Carrie Furnaces, he replied that the chances were as good as any.

Ron is the Director of Historic Resources and Facilities at the Rivers of Steel National Heritage Area, where he works to preserve and re-activate the area's industrial structures. He was growing up in Pittsburgh when Homestead shut down, so he has experienced as much of the decline of the industry as its afterlife. But many years on the site, in the archives, and with former steel workers who are now active in the Heritage Area have given him a visceral understanding of that history. When he began at Rivers of Steel, a private development company had purchased the site and planned to dismantle the

Figure 3.9 Ron Baraff in the stationary car dumper where material loads were received, tipped from train cars, and transferred mechanically to the ore yard, Carrie Furnaces, 2014

complex for scrap metal. By the time there was enough political momentum to save the structures, all but Furnaces 6 and 7 had been torn down. Meanwhile, all of the other blast furnaces and most other major steel industrial structures in the area had also been demolished. Now, Furnaces 6 and 7 remain as rare and comprehensive examples of pre–World War II integrated smelting technology in the region, gaining recognition as a National Historic Landmark site.[27]

When Andrew Carnegie purchased the Carrie Furnaces in 1898, he added another strategic node to Carnegie Steel's growing empire of holdings. Without a continual and immediate source of iron, Carnegie's Homestead mills had been transporting iron from blast furnaces miles down river and were vulnerable to infrastructure disruptions.[28] The Carrie Furnaces would provide a dedicated supply of iron to Homestead, and together they would evolve into one of the company's flagship integrated plants. Carnegie envisioned a business model free from dependencies and strengthened by market dominance. The company gained control of ore and fuel supplies, rail networks, and production facilities; modernized and expanded facilities; and made powerful alliances to expand the company's market reach. Late-nineteenth-century iron and steel companies generally followed these principles, as the move to consolidate and modernize could increase plant efficiencies, lower costs through scale economies, and undersell the competition – and produce larger and more powerful agglomerations or "trusts."[29]

The ultimate consolidation of power – the United States Steel Corporation – was realized in 1901, when J.P. Morgan and Elbert Gary purchased the Carnegie Steel Company and merged it with two of its major competitors, the Federal Steel Company and National Steel Company, as well as five major specialty product firms. Together they amassed 149 steel works, 87 blast furnaces, and 1,000 miles of railroad, as well as vast coking coal, limestone, and iron-ore reserves;[30] they comprised 65 percent of the national steel market, and had a capitalization that far exceeded the federal government's budget.[31] By dominating the market, U.S. Steel was able to set prices and control the industry at large.[32] So large that it was known simply as "the Corporation," U.S. Steel demonstrates the strategies of monopoly capitalism and its ability to reorganize huge expanses of land and enormous labor resources.

Between Carrie Furnaces 6 and 7 and the train tracks that line the river lies the former ore yard, almost a couple of football fields long and one wide. Here, mountains of iron ore, coke, and limestone – the precursors of pure iron – arrived on rail lines and at barge docks that flanked the Monongahela. As we walk, Ron searches for and then picks up a few granules from the gravelly ground and drops them into my open palm. One is round, hard, and brownish-grey – iron ore;[33] one is lightweight and puckered black – coke; and one is chalky and brittle – limestone; and from another part of the yard, one is aerated and matte – the slag skimmed off the top of molten iron.

Of the materials in my hand, the iron ore had traveled the farthest. The metal from Minnesota's Mesabi Iron Range, a 110-mile-long iron-rich ribbon just east of Lake Superior, had settled there 2 billion years before. Seeping up from underwater volcanoes and sea-floor cracks, dissolved iron saturated

Figure 3.10 **Iron ore at Carrie Furnaces, 2014**

the earth's primordial oceans until cyanobacteria populations proliferated. As they did, the photosynthetic cyanobacteria released spectacular amounts of dioxygen into the earth's no-oxygen environment. Dissolved red hematite and magnetite bound and oxidized, or rusted, with this new oxygen, settling on the ocean floor, alternating with iron-poor strata of shale and chart in graphic layers known as Banded Iron Formations. Geologist William Stropf calls this dramatic process "the rusting of the earth."[34]

Where Banded Iron Formations ended up near the earth's surface, speculators saw a profitable material resource. They were close to the Great Lakes transportation network, and therefore to Pittsburgh and Chicago, and so the Mesabi Range deposits became a lynchpin of twentieth-century industrialization.[35] For Carnegie Steel, and then U.S. Steel, investments in the Oliver Mining Company and Minnesota ore lands secured material sources and carried the Corporation, providing three-quarters of its ore by the early twentieth century,[36] and a billion tons of ore to the nation by the time that Riverside Park was completed.[37]

One could witness the magnitude of U.S. Steel's extraction in the population growth of Hibbing, Minnesota (which increased six-fold in the first two decades of the twentieth century), and also in the gaping hole of its mining operations. By the 1920s, when U.S. Steel merged three of its mines to create the Hull-Rust-Mahoning conglomerate open-pit mine, it had swallowed the land that Hibbing stood on, forcing a relocation of the entire town.[38] Unlike the labor and equipment required for underground mining, surface iron deposits could be extracted using locomotives and steam shovels, with comparatively fewer skilled workers and at a much lower cost.[39] With each blast of dynamite, workers ran for cover as ore was freed from the earth. Steam shovels then scooped up 15 tons at a time into ore cars, and locomotives zig-zagged their way up switchbacked rail tracks out of the steep mine walls, pushing through the haze of their own smoke.[40]

After securing a bounty of cheap, high-quality iron ore, the critical next step in integration was to improve and control the transportation infrastructure.

Figure 3.11 **Rochleau Mine, Virginia, Minnesota, 1947**

Source: Photograph by Zweifel-Roleff Studio (Zweifel-Roleff Studio Photograph Collection for DM&IR and Oliver Iron Mining, 1942–1954, S3744, Archives and Special Collections, Martin Library, University of Minnesota Duluth).

While the Mesabi mines were located conveniently close to the Great Lakes inland waterway, both the pits and lake transport couldn't operate during the winter months, raising the stakes for maximum material supply during the warmer months. Carnegie Steel quickly identified points of weakness for greater control. To float ore from the port at Duluth through Lakes Superior, Huron, and Erie, they established the Pittsburgh Steamship Company; to modernize the Lake Erie docks at Connaught, Ohio, they purchased and overhauled them; to bypass the Pennsylvania Railroad's fees and get ore from Connaught to Pittsburgh, they purchased an old rail line, and then rebuilt and redirected it to pass through Pittsburgh and connect with the Union Rail lines, which linked all of its riverfront works.[41]

From Connellsville, Pennsylvania, came the coke. Southwestern Pennsylvania is streaked with coal deposits, and the thick, soft, and accessible Connellsville seam was an ideal source for metallurgical coke, which was becoming the primary fuel of the nation's blast furnaces and steel mills. When heated in the absence of oxygen, coal releases volatile gases and transforms into coke, a hard and porous material rich in combustible carbon. Coke burns hotter than coal and is able to melt iron ore as well as flux. Under heat, coke serves as a reducing agent, aiding iron oxides to transform to more purified elemental iron. As a structural material, it also supports the weight of iron ore in the blast furnace, allowing for the circulation of air between its piled-up fragments.[42]

Carnegie had partnered with Henry C. Frick in the late nineteenth century, gaining joint control of two-thirds of the region's coking ovens and vast coal reserves. By World War I, the traditional dome-shaped beehive coking ovens, typically located near coalmines, were abandoned for new byproduct coke ovens that recycled gases, oil, ammonia, and tar from the coking process for use within blast furnaces and steel mills.[43] For U.S. Steel in Pittsburgh, this meant building the Clairton Coke Works just five miles down the Monongahela from Homestead, a new byproduct coke plant that could supply coke and gas via pipeline to fuel Homestead and its other steel works on the river's edge. In a single plant, Clairton was able to produce 40 percent as much coke as had been produced throughout the entire Connellsville region at its peak.[44]

Limestone, the third precursor of iron making, serves as *flux* (Latin for flow), a reducing agent, by drawing out impurities from the iron ore, binding with them, and becoming slag. The slag is later disposed of or recycled for other industrial uses. As with ore and coking coal, Carnegie Steel and later U.S. Steel kept close tabs on limestone sources. By the turn of the century, Carnegie owned 75 percent of the capital stock of the Pittsburgh Limestone Company,

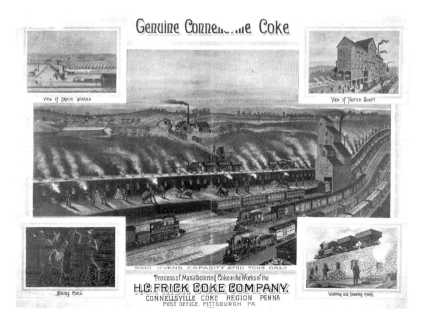

Figure 3.12 **Process of manufacturing coke at the works of the H.C. Frick Coke Company, Connellsville Coke Region, Pennsylvania**

Source: Library of Congress Prints and Photographs Division, www.loc.gov/item/2003673029/.

Figure 3.13 **Pittsburgh Limestone Company quarry, Ganister, Blair County, 1922**
Source: Photograph by Ralph Walter Stone (Pennsylvania Geological Survey).

and other future U.S. Steel subsidiaries operated quarries in nearby Blair, Butler, and Lawrence counties.[45]

From the Mesabi ore lands to the Connellsville coalmines to Blair county limestone quarries, the project of integration reorganized vast geological deposits, siphoning them towards blast furnaces and steel mills. All of the mineral land acquisitions, rail line realignments, port construction, and steamship purchases reflect how maximization of profits is tied to geography: the locations of certain geological materials with specific properties, their proximity to natural waterways and engineered routes, and the weight of iron ore and its effect on freight costs.

As we climb the stairs leading up to the ore bridge, Ron narrates the abandoned structures as if they were in full production mode. Ore, coke, and limestone arriving by rail and river would be mechanically transferred into the ore yard. While today it is a flat, grassy expanse, in the 1930s the ore yard contained ore piles that dwarfed the trains below. On the outside of the furnaces, skip cars carrying iron ore, limestone, and coke ascend like funiculars on a steel cliff, and then tip into a holding hopper suspended over top of the furnace. In cycles, the materials are released into the furnace shaft, piling up into a massive column that gradually descends as it morphs. Hot air from coke-fueled furnaces blasts from below, igniting material at the bottom and setting off a series of chemical reactions that unfold vertically. With the hot blast, the coke's carbon gasifies to carbon monoxide, which, rising through the material stack, reduces the iron oxides and releases carbon dioxide. As descending ore and limestone confront the 2,000°C air, the pure iron melts and collects at the bottom of the furnace, and the limestone binds with non-metallic minerals into a foamy slag that floats on top. Ron shows me how workers would tap the furnace, releasing flumes of slag into cars directed to nearby Brown's Dump and

Figure 3.14 Ore yard, Carrie Furnaces, circa 1920–1940
Source: Photograph by William J. Gaughan (University of Pittsburgh Archives and Special Collections).

the molten iron into ladle cars. The latter headed south across the Hot Metal Bridge over the river, towards the Homestead steel mill.

At the mill in large open-hearth furnaces at extremely high temperatures, the molten iron is combined with scrap steel, limestone, and other additives, and when finished, cast from overhead cranes into heavy, rounded ingot molds. White-hot ingots of steel pass then through smaller and smaller rollers, with each pass elongating and taking shape as slab and then plate, or as angle, I-beam, or H-beam. From Homestead, orders were shipped to manufacturing plants like the American Bridge Company in Ambridge, just down the Ohio River, where they would be cut, welded, assembled, and finished according to contract specifications. U.S. Steel's American Bridge Company fabricated and installed structural steel for some of the nation's most triumphant structures in the early twentieth century, including the Empire State Building, the Chrysler Building, and the Henry Hudson Bridge in New York City alone.[46] And, in the mid-1930s, American Bridge would also supply steel to restructure the Hudson River shoreline, to cover New York Central rail tracks, and to construct the West Side Elevated Highway.[47]

Figure 3.15 (Clockwise from top left) (A) Molten iron from the Carrie Furnaces is poured into one of the hot metal mixers in the No. 5 open-hearth shop of U.S. Steel's Homestead Works; (B) Steelworker gauges the thickness of white hot steel plate emerging from the rollers of the 160-inch plate mill at U.S. Steel's Homestead Works (Photograph by and © Fritz Henle); (C) Steel beams are removed from the storage yard for shipment from the structural mill at Homestead Works (Photograph by and © Fritz Henle); (D) Loading barges on the Monongahela River at the central loading dock of U.S. Steel's Homestead Works, bound for the American Bridge Company Plant for fabrication

Source: United States Steel Corporation Photographs, Volume 21. Baker Library, Harvard Business School.

Riverside Park, 1937

"Speaking of structural steel," former U.S. Steel president Charles Schwab remarked in 1926, "where has there been such a use of it as we have seen in New York City and other points on the Atlantic seaboard in this present wave of new construction?"[48] In the first two decades of the twentieth century, steel production increased by 400 percent, and steel for construction increased by 250 percent, as large office buildings shot up in Chicago, and then, increasingly, New York City.[49] By the late 1920s, the New York metropolitan area consumed more than one-quarter of the country's national production, or 570,000 tons of structural steel per year.[50] In that decade, New York saw more tall buildings constructed than at any other time in the city's history,[51] and preliminary work on infrastructural improvements along the Hudson River, including in Riverside Park, were all planned in steel.

As economists William N. Goetzmann and Frank Newman suggest, the proliferation of towers was a financial rather than an architectural phenomenon. For the first time, the main purpose of new construction was not to house corporate headquarters, but to maximize rents and developer profit through real estate speculation;[52] a new tower was, as architect Cass Gilbert said, "a machine to make the land pay."[53] Laissez-faire building codes, high returns on real estate securities, and a lack of financial regulation sent the skyline soaring. Steel towers were not only a *place* of business, Carol Willis writes in her book *Form Follows Finance*, they *were* a business.[54] While Wall Street solidified as a global financial center, industrialists and developers amassed and multiplied their fortunes in steel towers that sprung up across Manhattan.

"The New York skyline," Goetzmann and Newman write, "is a stark reminder of securitization's ability to connect capital from a speculative public to building ventures . . . Optimism in financial markets has the power to raise steel, but it does not make a building pay."[55] And the buildings didn't pay. This cocktail of speculation and investment in public securities helped sink the stock market in October of 1929. Attempting to quell the panic just after the crash, the vice president of the New York Stock Exchange placed a bid for a large quantity of above-market U.S. Steel shares – a gesture that failed to reverse the market's downward spiral.[56] And as wages, employment, agriculture, and heavy industry plummeted with it, construction projects – including longstanding plans to cover the New York Central rail tracks in Riverside Park – screeched to a halt.

As devastating unemployment, widespread hunger, and social protest mounted in the years following the crash, President Franklin Delano Roosevelt's New Deal administration responded with emergency relief programs designed to create jobs and stimulate industrial production. The federal government, leaning on the pre-existing capacities of municipal agencies (while also expanding their scope), sought to employ as many people as possible, while building an ambitious array of public works. Workers would build material intensive infrastructures through the Public Works Administration (PWA), such as bridges and dams, and, through the Works Progress Administration (WPA), public projects, such as highways and parks. At this time, American production, in stark contrast to before, was directed not towards private consumption, but towards public service.[57]

New York City – with its swelling unemployed population, immense political power, and effective organizational capacity – fared well in these national programs, receiving a disproportionately high share of the nation's WPA funding.[58] Parks Commissioner Robert Moses brokered these emergency funds (along with other grants, debts, and deals) and redirected them towards gargantuan visions. Through a particular mixture of administrative genius, municipal support, New Deal funding for labor and materials from the PWA and the Civil Works Administration (CWA), and racially discriminatory cost-saving decisions, Moses mobilized the rehabilitation of Riverside Park as part of the West Side Improvement, the largest construction project in the city's history.[59] The ten-and-a-half-mile West Side Improvement strung together the elevated rail in Lower West Manhattan, the West Side Elevated Highway, and the Henry Hudson Parkway that started at 72nd Street and ran through Riverside Park up to Van Cortland Park in the Bronx. Under the guise of a single, enormous public work, the West Side Improvement reorganized rail, vehicular, and pedestrian networks and recreation space along the Hudson River, resolving conflicts in Riverside Park between trains and park-goers, and in lower Manhattan between at-grade trains and pedestrians (especially at 11th, the so-called "Death Avenue").

After decades of failed proposals and disputes, the conflicted and disjointed landscape of Riverside Park would be reshaped into a fluid assembly

Figure 3.16 **Riverside Park under construction, showing steel frames installed over New York Central Railroad tracks, view looking south from 82nd Street, Hudson River beyond, 1936**

Source: Milstein Division, The New York Public Library.

Figure 3.17 **Riverside Park construction, rail tracks have been covered, view looking north from 72nd Street, September 15, 1937**

Source: Milstein Division, The New York Public Library.

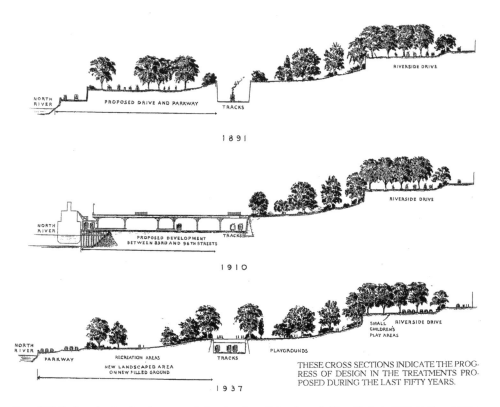

Figure 3.18 Sections showing changing proposals for mitigating rail disturbance in Riverside Park, 1891, 1910, and 1937

Source: Harry Sweeny, West Side Improvement, Published on the Occasion of the Opening, October 12th, 1937. New York: Department of Parks, 1937, p. 15.

of recreation, highway, and rail. The roofing of the rail tracks, expansion and filling of the shoreline, smoothing of the grade between Riverside Drive and the Hudson River, and integration of parkway and promenade would produce fifty-three new hectares of parkland valued at $23.7 million.[60] In 1934, a bill signed by Senator John Buckley authorized funding for relief workers to begin roofing the New York Central Tracks in Riverside Park.[61] To the west of the tracks, the large encampments of people living on the dumps were evicted, material storage lots were cleared away, and large tracts of submerged land were filled in with subway excavate to extend and straighten the shoreline.[62] By January 1936, "literally acres of steel and concrete"[63] had been installed, and 5,800 tons of structural steel had been ordered from the American Bridge Company to cover New York Central's tracks.[64]

Steelworkers installed pre-fabricated, rigid steel frames made of plates and angles, riveted together in intricate patterns, all along the rivers edge.[65] Spanning more than 76 feet wide, these frames were erected every 17 feet along the length of the tracks in a rhythmic repetition, varying with local site conditions along the line.[66] Where rail lines diverged, a row of columns filled in for support, with X-shaped cross-braces marking columns and their girders above. Perpendicular to the frames and parallel to the tracks, stringers were secured to the frame girders at 5-foot intervals, riveted to stiffener angles or supported on and bolted to shelf angles, all tying the innumerable pieces into one.[67] Fastened to, hanging from, or mounted on top were all forms of other steel castings and fittings: conduits, boxes for subsurface drains, gutter covers, expansion joint plates, lamp standards. The structure's east side attached to the existing retaining wall, and on the west side it was anchored to the ground, sometimes adorned with stone, other times completely obscured by the slope.

Figure 3.19 **West Side Improvement, steel framework, 72nd St., looking east, 1935**

Source: New York City Parks Photo Archive.

Figure 3.20 **Steel workers assembling the West Side Highway as part of the West Side Improvement, 1937**

Source: Photograph by Samuel H. Gottscho (Gottscho-Schleisner, Inc./Museum of the City of New York, 88.1.1.4698).

Once steel roofing of a section was complete, highway construction, topography, and planting followed. Above the structure, steel-reinforced concrete slabs were topped with waterproofing and various surface materials. In the newly structured promenade and adjacent parklands, soil was installed and prepared for the planting of turf, trees, and shrubs. A photograph commemorating the park's opening shows the dark rail tunnel with a diagonal shaft of light piercing through. Celebrating the project's infrastructural triumph – separating the train tracks from the park above – the caption reads, "Above these covered tracks are grass, trees, and sunlight."[68] While appearing as a continuous landmass, this new sinuous landscape was ultimately supported by 45,000 metric tons of structural steel.[69]

This steel-structured landscape offered a precedent for the integration of highway and parkland, and asserted a new form of active public landscape. The final design, overseen by consulting landscape architect Gilmore Clarke and architect-engineer Clinton Lloyd, was a truly modern landscape. By combining highway and recreation agendas, it reflected the type of landscape project privileged by New Deal public works funding. In 1937, the same year that Riverside Park was completed, two-thirds of government grants for non-federal public construction supported road construction and recreation facilities, perhaps

Figure 3.21 Steel structure covering Riverside Park's rail tracks in 1936, and with landscape construction complete in 1937

Source: Sweeny, West Side Improvement, p. 21.

Figure 3.22 "Above these covered tracks are grass, trees, and sunlight."

Source: Sweeny, West Side Improvement, p. 49.

the most visible of the public landscape (the rest included public water systems, sewerage systems, public buildings, and other transportation facilities).[70] Clarke's firm, Clarke & Rapuano, established with Michael Rapuano in 1934, represented a new type of landscape architect firm – one in service of public works, and one that wielded engineering to achieve its aesthetic ends.

Geographer Matthew Gandy writes of how this generation of parkway design (including the Taconic Parkway and the Bronx Parkway, other projects that Clarke worked on) both offered a new technological sublime and a new regional imaginary of nature.[71] These projects represented a privileged means for escaping the difficulties of the city and consuming the exurban as a landscape of leisure.[72] While on the one hand Riverside Park's redesign hid and separated rail traffic from the park experience, it also foregrounded and celebrated modern mobility. The *pièce de résistance* was the 79th Street Rotunda. The park's focus was not a central allée nor formal garden, but a heroic traffic feature complete with restaurant, café, artworks, and a jubilant fountain in its center. When critics charged the Riverside Park renovation with obstructing the river's edge with the highway, Clarke argued that park visitors could see the river overtop of the parkway, and dismissed the idea that direct contact with the river would ever be desirable – it was and would stay, after all, an "open sewer."[73] For the first time, the park's aesthetics were also described from the driver's seat.

"This is not the kind of romanticism one finds in Olmsted's coyly naturalistic parks," architectural critic Lewis Mumford wrote in praise of the renovation,

Figure 3.23 **79th Street rotunda, 1937**

Source: Photograph by Samuel H. Gottscho (Gottscho-Schleisner, Inc./Museum of the City of New York, 88.1.1.4653).

later calling it "thoroughly modern planning." Mumford applauded the park's decisive lines and curves, drawn to facilitate the speed of modern citizens. Far from a refuge for solitary nature-lovers, this was a bold, daring park design to accommodate a vast public in a "grand collective design."[74] The design introduced a whole spectrum of active recreational facilities as part of the West Side Improvement, including seventeen playgrounds, five football fields, eighteen horseshoe courts, twenty-two tennis courts, and extensive bicycle paths.[75]

The renovations to Riverside Park (and other public works built in the same era) reorganized the public realm, transplanting social activities that used to take place in the streets and formalized them in park facilities. And for citizens unable to escape the city on its scenic parkways, it also gave them more access to active recreational opportunities that hadn't previously existed.[76] But in Riverside Park, these amenities and improvements were not equitably distributed. Robert Moses made an early decision to prioritize park spending between 72nd Street and 110th Street – where the wealthiest and whitest populations were most likely to be – and to spend one-quarter as much per mile north of 125th Street, where the population was noticeably poorer and where African American communities were concentrated. Through this discriminatory calculation, he saved nearly $30 million.[77] Between 125th Street and 155th Street, the tracks remained uncovered, no increase in parkland was made, commercial warehouses remained, and scant trees and plantings were installed. In that area, just as at the turn of the century, residents continued to be subjected to the smell of live cattle and the non-stop racket of the railroad, still open to the air.[78]

The renovations to Riverside Park wielded New Deal funding and labor, the sectional arts of landscape architecture, and the structural capacity of steel to choreograph rail, pedestrian, and vehicle movement into a highly constructed, connective landscape. Where private construction had dwindled during the depression, public construction stepped in to boost. The West Side Improvement and scores of other massive public works projects not only put thousands of people to work, but they also consumed huge quantities of construction materials – including steel.

U.S. Steel, Pittsburgh, 1937

As New Deal programs injected funding into public infrastructure projects, consequentially stimulating the steel industry, other New Deal legislation directly supported the steel industry and, most profoundly, its workers. To stimulate economic recovery for steel and the nation's other faltering industries, Congress passed the 1933 National Industrial Recovery Act (NIRA), which authorized government regulation of industry and allowed companies to set prices. While big steel generally approved of the NIRA, they rejected Section 7a, a clause that guaranteed the right to organize and protect collective bargaining rights for unions. The Supreme Court soon declared the NIRA unconstitutional; however, the subsequent Wagner Act of 1935 solidified labor's power, outlawing company-controlled unions and anti-union activities, legalizing union organizing, and establishing the National Relations Labor Board to oversee these new regulations.[79]

This legislation shattered the powerful fraternity of government, business, and the police force, which had long contested labor.[80] From the Homestead Lockout of 1892, to a massive steelworker strike of 1919, large industrial corporations like U.S. Steel were used to unconstrained support from local, state, and federal government, as well as the police.[81] Their profits, drawn by cutting costs and increasing production, fundamentally relied upon suppressing unions, and they had the unbridled power to do so.[82] As part of a systematic campaign of repression, U.S. Steel employed private police, espionage, harassment, and blacklists to weed out and intimidate organizers. In addition, the company exploited racial tensions to divide workers, sorting workers by race and ethnicity, and fueling anti-black and immigrant hostilities by threatening white workers with replacement by African Americans or European immigrants.[83]

Buoyed by Section 7a and the subsequent Wagner Act, workers of all stripes, from coppersmiths and hotel workers to mechanics and miners, began to organize, and a wave of strikes swept the nation in 1933.[84] Finally able to mobilize at a national scale, movements brought the plight of industrial workers to the public eye. Long plagued by exclusionary practices based on race or skill level, unions moved towards the notion of industrial unionism – the ideal of organizing across skill-level, race, and ethnicity. To this end, after successfully organizing coal on an industrial basis, the leadership of the United Mine Workers of America (UMWA) founded the Congress of Industrial Organizations (CIO) and several industry-specific unions, including the Steel Workers Organizing Committee (SWOC). While the SWOC remained one of the more hierarchical and conservative unions, with the help of the National Association for the Advancement of Colored People (NAACP), black churches, and community leaders, it began to grow its African American membership, as well as the membership of ethnic groups that had traditionally been excluded.[85] As CIO membership ballooned, industrial unionism took hold as a significant new institution of labor power.[86]

In March 1937, the same year that Riverside Park's renovations were completed, U.S. Steel Chairman Myron Taylor shocked the nation by signing an agreement with the SWOC, what historian John H. Hinshaw describes as "one of the most important incidents in the history of modern labor."[87] The agreement recognized full union representation with SWOC as the agent, and established a five-dollar hourly wage, an eight-hour workday, a forty-hour workweek,[88] and grievance procedures.[89] Although broadly understood as a massive victory for labor, it benefited U.S. Steel in other ways: the agreement tied steel prices to wage increases, offsetting labor gains to consumers and setting a precedent that would ultimately generate inflation.[90]

U.S. Steel's capitulation to SWOC – the first such agreement by a major American industrial corporation – indicated the growing power of industrial unionism, as well as the government's significant role within it. Sit-down strikes had crippled General Motors just months before, demonstrating the stakes and potential losses that corporations faced by not negotiating. With business rebounding, U.S. Steel recognized how damaging a strike could be. But beyond the threat of strike, the agreement signaled U.S. Steel's reliance on an increasingly pro-labor government and, in particular, on governmental military contracts. By withholding profitable armament contracts, the Labor Department compelled U.S. Steel to negotiate with the unions. With war on the horizon, cooperation with the government was more valuable to big steel than any perceived loss to labor.[91]

The simultaneous completion of Riverside Park (and the West Side Improvement) and the signing of the U.S. Steel–SWOC labor agreement reflect a critical moment in which power struggles between industrial corporations and the public were reorganized – physically and socially – through steel. In Riverside Park, relief funds and labor built a modern new landscape, staking a claim for the public (one with significant exclusions) and separating incompatible industrial and civic land uses through the tectonic shaping of landform. In Pittsburgh, the public stakes were made clear less through landscape and more through the social organization of the local workforce. The unionization of U.S. Steel transformed the power dynamics of the steel industry at large, challenging the corporation's omnipotence and asserting the worker's place at the bargaining table. In both cases, New Deal policies, responding to protracted struggles and the threat of social unrest, enabled these shifts to occur.

The two riverside sites had different fates. New York City reaped the benefits of New Deal emergency-funded projects, such as Riverside Park, which addressed health and environmental concerns of a more affluent, white population living close to industrial activities. Allegheny County saw a smattering of WPA-funded landscape and infrastructural improvements in the late 1930s, including a new steel bridge spanning the Monongahela River and connecting the Homestead Steel Works. Yet, as industrial workers, Pittsburgh residents had a complicated relationship to environmental remediation and regulation. Bearing the brunt of poor air and water quality, workers associated clear skies with idle factories and coal plants, and smoke with employment and prosperity.[92]

The Third Century

After the tunnel's completion in 1937, the New York Central line carried freight under Riverside Park visitors for decades. Rail traffic slowed during the 1960s as larger and better-connected rail yards in the Bronx and New Jersey were constructed, until a decade later when it finally stopped. By the mid-1970s, people started moving in. With a steel deck, solid concrete walls, and few access points, the tunnel under Riverside Park was a safe place for people seeking shelter. Just as New Yorkers built encampments along the shores of Riverside Park during the Great Depression, people set up house in the tunnel as homelessness soared in the 1980s. As industry left New York City, moving south and westward, the economic base shifted from manufacturing jobs towards lower-paying service industry jobs, impoverishing many. The combination of eroding federal aid, austerity measures, and the dissolution of social services worsened the effects of de-industrialization for many, creating a newly visible homeless population.[93] Accounts from the early 1980s describe how homeless men could now readily be seen in many parts of the city, often occupying abandoned freight yards and unused rail lines.[94]

Nicknamed the Freedom Tunnel after one of the many graffiti artists (Chris "Freedom" Pape) working in it, the tunnel under Riverside Park became home to a longstanding community of residents, documented in Marc Singer's film *Dark Days* and Margaret Morton's photo-book, *The Tunnel: The Underground Homeless of New York City*.[95] In one of Morton's photographs, a shaft of daylight from Riverside Park illuminates the ground below and recalls the 1937 photo mentioned earlier in this chapter. In the tunnel foreground, a small

Figure 3.24 **Bernard's Tree, 1995**

Source: Photograph © Margaret Morton (*The Tunnel: The Underground Homeless of New York City*. New Haven: Yale University Press, 1995).

pile of soil is scattered with dried leaves and a small tree sticks upright.[96] It isn't clear whether someone has stuck a branch into the soil, or if it has taken root there on its own. One of the tunnel residents tells Morton, "Once the weather really breaks and gets warm, certain seeds will drop through the grate from up top and things sprout over there . . . That area has always had plant life."[97] While invisible from the park above, the tunnel slowly accumulated soil, leaf and human litter and seeds through steel gratings, sometimes sprouting pieces of the park down below. As Amtrak prepared to install a new commuter line through the tunnel in 1991, residents of the tunnel were evicted, and eventually, trains passed through it again.

While the steel industry supplied World War II and therefore thrived, American steel production fell in the second half of the century; between 1970 and 1989 production dropped by 25 percent and steel employment dropped by 59 percent.[98] During this time material substitutes like aluminum and polymers began to encroach on steel's market share, foreign imports rose by 50 percent,[99] and in Pittsburgh, initial geographic advantages gave way as steel production shifted to California, Utah, Houston, and Alabama.[100] With deindustrialization, Pittsburgh's workers suffered, facing sustained and bleak prospects of poverty and unemployment, with black communities faring the worse.[101] Deindustrialization in Pittsburgh is sometimes chalked up to mismanagement or poor union relations, but historian John H. Hinshaw argues that Pittsburgh's industrial collapse was instead the normal functioning of capitalism:

> deindustrialization was less a failure on industrialists' part than a strategy . . . Steel was simply a means to an end. Industrialists amassed spectacular profits from steel for many years, but when it became more profitable for steelmakers not to make steel, they invested elsewhere.[102]

The flight of the steel industry from Pittsburgh left behind a constellation of massive brownfield sites, each heavy with the specter of lost jobs. Of those tied to U.S. Steel's Homestead integrated steel works, two are now vast retail environments and one, the Carrie Blast Furnace site, is still in development. Brown's Dump, where molten slag from Carrie was dumped for decades, became a new landform of the local landscape. When dumping stopped, the hill became a profitable development site for Century III, a mega-shopping complex named to inaugurate the country's new third century. Today, the mall stands on the brink of abandonment; its empty corridors and shuttered flagship stores a melancholic vision of the century that lies ahead. One can drive up the massive slag deposit, first to the empty Century III parking lot, and then even higher to the more active Home Depot close to the top, with panoramic views of the surrounding hilltops. The Homestead Steel Mill site, once one of the world's largest steel mills, became The Waterfront, an open-air shopping center, in 1999. In 1988, two years after U.S. Steel shut down the Homestead Steel Mill, a development corporation bought the site and razed the hundreds of buildings that filled the 430-acre site, selling their metal as scrap.[103] Most remnants of the steel mills are gone except for the single row of brick smokestacks

Figure 3.25 **Slag deposit of the former Brown's Dump at Century III Mall parking lot, West Mifflin, 2015**

that line the southern parking lots and serve as the shopping center's logo. As mentioned earlier, the Carrie Furnace site was also leveled and sold for scrap metal after closure, but the two remaining furnaces and the acres they sit on are now a major development project in the works.

While the Carrie Furnaces are unique in their own right, this type of site – contaminated, waterfront location with obsolete industrial structures – has become ubiquitous in North America.[104] Such sites have become central subjects for landscape architecture projects and theory of the late twentieth and early twenty-first century. Niall Kirkwood's 2001 book *Manufactured Landscapes: Rethinking the Post-Industrial Landscape* established the centrality of derelict industrial sites and the technology of remediation to the discipline.[105] At Gasworks Park in Seattle, designed by landscape architect Richard Haag, visitors sprawl and sunbathe on manicured lawns between the behemoth coal gasification structures. Completed in 1975, the park brought the public in close proximity to obsolete industrial structures and incorporated bioremediation strategies, challenging assumptions of what a leisure park might look like. In the 1990s, Landscape Park Duisburg Nord in Germany's Ruhr Valley designed by Latz + Partner, took public engagement with industrial structures to the next level, sending people climbing up them and scuba diving into converted tanks. The design cultivated an industrial aesthetic, an appreciation for weathered steel and spontaneous vegetation, and a corporeal experience of the scale of the structures.

These are just two influential projects of countless others occupying similar sites – sites that share an economic (and often legal) imperative to remediate contaminated land, and to develop for a new post-industrial user. From a theoretical perspective, these projects are fascinating because in addition to introducing innovative technologies for dealing with contaminated land and water, they allow the public to confront the industrial history and toxic legacies of these sites. But as landscape theorist Elizabeth Meyer asks, how does this fascination resonate for the communities associated with the formerly industrial land and living next to this new park type that contains latent risks?[106] Furthermore, confronting industrial history through recreation as a professional urbanite is quite different than as a former steel worker with a longstanding relationship to both the benefits and risk of that industry. In addition to remediation and providing recreational access to new publics, how might this type of project confront the evacuation of industry and the economic devastation and locally experienced environmental risks that were left in its wake?

In contrast to Brown's Dump or the Homestead Steel Mill site, the development of the Carrie Furnace site has been drawn out, but this slowness may have been an asset. Neighboring communities of Edgewood, Rankin, and Swissvale contributed to a Comprehensive Plan for the site, asserting the need for local workforce development, sustainable manufacturing, green building practices, and open space preservation that integrates the historic structures.[107] Erin Deasy from Allegheny County Economic Development says that communities wanted working-class jobs, not minimum-wage jobs like at the nearby malls. Malls are a shell game, she explains, you put one mall here and when another one gets built down the road, all the business can leave just as soon as it came.[108] A call for development proposals offers 110 acres of developable land, currently undergoing remediation, on either side of the furnaces.

Rivers of Steel will continue to oversee the historic Carrie Furnace structures, offering programs that include tours led by former steel workers, film festivals, and iron-casting workshops. When Ron Baraff began at Rivers of Steel, he says that some people disliked the graffiti and vegetation that had overtaken the site in the decades since it shut down. But Ron believes that the graffiti and artist culture that developed around the closure of the plants is relevant to the site's history, and Rivers of Steel has made an effort to reach out to artists rather than erase their work. The spontaneous vegetation that emerged alongside the steel industry and in its aftermath is also an element to work with rather than to remove. In an ongoing project titled *Addition by Reduction*, writer and horticulturalist Rick Darke selectively removes spontaneous vegetation, shaping the site over time to support public use and understanding of the site's labor, industrial, and ecological history.[109] Ron tells me that, in general, Rivers of Steel is hesitant to change too much of the Carrie Furnace site, aware of the ways that history can be reduced and made token, and taking care not to sanitize or manicure it. A major renovation of the furnace site would cost too much anyways, so for now they are taking their time. Perhaps the commitment to local economic development, working-class employment, and ecological regeneration can

Figure 3.26 Rick Darke's ongoing landscape installation, Addition by Reduction, Carrie Furnaces, 2014
Source: Photograph by Rick Darke.

foster a new kind of "post-industrial" development, one that creates socially and ecologically supportive forms of work that are truly worthy of the third century.

Notes

1 Robert E. Henshaw, *Environmental History of the Hudson River: Human Uses That Changed the Ecology, Ecology that Changed Human Uses* (Albany: State University of New York Press, 2011), xxiv.
2 Riverside Park South is constructed on the former 60th Street Yard of the New York Central Railroad, abandoned in 1976. Thomas Balsley & Associates led the design team for the park, whose first phase opened in 2001.
3 Jens Jensen, *Report to the League for the Protection of Riverside Park, 29 November 1916* (New York: Women's League for the Protection of Riverside Park, 1916), 2.
4 The Hudson River Railroad initially established tracks along the river in 1846.
5 Harry Sweeny, *West Side Improvement: Published on the Occasion of the Opening October 12th, 1937* (New York: Moore Press, 1937), 18.
6 Ibid., 14.
7 Edward Grimm, *Riverside Park: The Splendid Sliver* (New York: Columbia University Press, 2007), 22.
8 New York Department of Public Parks, *Third General Report of the Board of Commissioners of the Department of Public Parks* (New York: William C. Bryant & Co., 1875), 300.

9   Ibid., 302; Riverside Avenue was renamed Riverside Drive in 1908. Grimm, *Riverside Park: The Splendid Silver*, 25.

10 Robert A.C. Smith, *The West Side Improvement and Its Relation to All of the Commerce of the Port of New York* (New York: M.B. Brown Printing & Binding Co., 1916), 80.

11 "West Side Tracks Plan is Complete," *New York Times*, April 8, 1916, 5.

12 Women's League for the Protection of Riverside Park, *Some Plain Facts about Riverside Park* (New York: Women's League for the Protection of Riverside Park, 1916), 1.

13 Robert L. Duffus, "Era of New Splendor Opens for West Side," *New York Times*, July 14, 1929, XX3.

14 Langdon White, "The Iron and Steel Industry of the Pittsburgh District," *Economic Geography* 4, no. 2 (April 1928): 127–9.

15 Edward K. Muller, "River City," in *Devastation and Renewal: An Environmental History of Pittsburgh and its Region*, ed. Joel A. Tarr (Pittsburgh: University of Pittsburgh Press, 2003), 54; Kenneth Warren, *The American Steel Industry 1850–1970: A Geographical Interpretation* (Pittsburgh: University of Pittsburgh Press, 1988), 134, 138.

16 White, "The Iron and Steel Industry," 129.

17 Muller, "River City," 55.

18 Joel A. Tarr and Terry F. Yosie, "Critical Decisions in Pittsburgh Water and Wastewater Treatment," in *Devastation and Renewal*, 70.

19 "Smog Linked to Pneumonia: Blanket of Smoke and Fog Responsible for Death Rate in City of Pittsburgh," *The Globe and Mail*, November 6, 1937, 36.

20 H.L. Mencken, "The Libido for the Ugly," in *Prejudices: Sixth Series* (New York: Alfred A. Knopf, 1927), 187.

21 Muller, "River City," 57.

22 Frederick Law Olmsted Jr., "Pittsburgh: Main Thoroughfares and the Down Town District," in *Report to the Pittsburgh Civic Commission* (Pittsburgh: Committee on City Planning, 1910), 19–21.

23 Ibid., 23–4.

24 Andrew S. McElwaine, "Slag in the Park," in *Devastation and Renewal*, 180.

25 Ron Baaff, Personal communication, July 10, 2014.

26 "Structural Steel Orders," *Wall Street Journal*, September 22, 1936, 3.

27 United States Department of the Interior, National Park Service, "Carrie Blast Furnaces Number 6 and 7," *National Historic Landmark Nomination*, United States Department of the Interior, National Park Service, accessed February 13, 2018, www.nps.gov/nhl/find/statelists/pa/CarrieBlast.pdf.

28 Warren, *The American Steel Industry 1850–1970*, 136.

29 Kenneth Warren, *Big Steel: The First Century of the United States Steel Corporation 1901–2001* (Pittsburgh: University of Pittsburgh Press, 2001), 8.

30 Carnegie-Illinois Steel Corporation, *Growing with America* (Pittsburgh: The Corporation, 1948).

31 Robert P. Rogers, *An Economic History of the American Steel Industry* (New York: Routledge, 2009), 42. The major steel producing and special products companies merged into the United States Steel Corporation in 1901.

32 John H. Hinshaw, *Steel and Steelworkers: Race and Class Struggle in Twentieth-Century Pittsburgh* (Albany: State University of New York Press, 2002), 31. For example, while it was evident that the industry would migrate west from Pittsburgh towards Chicago to open to western markets, U.S. Steel had invested heavily in Pittsburgh. From 1907 to 1921, U.S. Steel imposed a so-called "Pittsburgh Plus" pricing system: anyone buying steel from anywhere in the country would need to pay freight as if it were coming from Pittsburgh.

33 This piece was likely taconite, a lower-grade ore used in the second half of the twentieth century.

34 William J. Schopf, *Cradle of Life: The Discovery of Earth's Earliest Fossils* (Princeton: Princeton University Press, 2001), 171.

35 Warren, *Big Steel*, 65.

36 Ibid.

37 "Iron Floats to Market," *Saturday Evening Post*, December 23, 1939, 33.

38 Warren, *Big Steel*, 65.

39 Edward P. Jennings, "The Mesabi Iron Range," *Science* 23, no. 575 (February 9, 1894): 73.

40 "Iron Floats to Market," 33.

41 Samuel Bostaph, *Andrew Carnegie: An Economic Biography* (Lanham: Rowman & Littlefield, 2017), 103–4.

42 Kenneth Warren, *Wealth, Waste, and Alienation: Growth and Decline in the Connellsville Coke Industry* (Pittsburgh: University of Pittsburgh Press, 2001), 1–2.

43 Ibid., 232–3.

44 Ibid., 237–8.

45 Orrin Hugh Baker, "The Development of the Steel Industry in America" (Thesis for B.Sc. in Mechanical Engineering, University of Illinois, 1907), 55, 61, 68, 79, 93.

46 "Contracts for Parkway Bridge Let 4 Hours After City Approves Plan," *New York Times*, June 15, 1935, 1.

47 "Structural Steel Orders," 3.

48 Warren, *The American Steel Industry*, 158.

49 Rogers, *An Economic History of the American Steel Industry*, 34, 37.

50 Warren, *The American Steel Industry*, 180.

51 William N. Goetzmann and Frank Newman, "Securitization in the 1920's," *NBER Working Paper Series* (Cambridge, MA: National Bureau of Economic Research, 2010), 3. https://doi.org/10.3386/w15650.

52 Ibid.

53 As quoted in Carol Willis, *Form Follows Finance: Skyscrapers and Skylines in New York and Chicago* (New York: Princeton Architectural Press, 1996), 19.

54 Ibid., 182.

55 Goetzmann and Newman, "Securitization in the 1920's," 19.

56 Warren, *Big Steel*, 144.

57 Mason B. Williams, *City of Ambition: FDR, LaGuardia, and the Making of Modern New York* (New York: WW Norton, 2014), 173.

58 Phoebe Cutler, *The Public Landscape of the New Deal* (New Haven: Yale University Press, 1985), 11.

59 Robert A. Caro, *The Power Broker: Robert Moses and the Fall of New York* (New York: Vintage, 1975), 527–33.

60 Sweeney, *West Side Improvement*, 4.

61 "West Side Project Ready to Hire 4,000," *New York Times*, May 28, 1934, 1.

62 Leonard H. Robbins, "Transforming the West Side," *New York Times*, June 3, 1934, XX2.

63 "Extending West Side Highway to the North," *New York Times*, January 5, 1936, XX6.

64 "Structural Steel Orders," 3.

65 Ambrose H. Stang, Martin Greenspan, and William R. Osgood, "Strength of a Riveted Steel Rigid Frame having Straight Flanges," U.S. Department of Commerce Research Paper RP1130, *Journal of Research of the National Bureau of Standards*, 21 (September 1938): 84780; New York Central Railroad, "Details of Frames 'Fla.b,' Roof Covering Over Tracks, W 98th St to W 111th St, New York City," August 1, 1936, Drawing 84780 (Olmsted Center, New York).

66 Historic American Engineering Record, "Henry Hudson Parkway, HAER No-NY 334," (Historic American Engineering Record, 2006), 65, https://riverdalenature.org/wp-content/uploads/2013/06/Historic-American-Engineering-Record-HAER-report-on-the-Henry-Hudson-Parkway.pdf.

67 New York Central Railroad, "Framing Plans and Details, Express Highway & Track Covering, W. 94th St. to W. 98th St., New York City," February 1, 1936, Drawing 84063 (Olmsted Center, New York).

68 Sweeney, *West Side Improvement*, 49.

69 Ibid., 20.

70 Cutler, *The Public Landscape of the New Deal*, 11.

71 Matthew Gandy, *Concrete and Clay: Reworking Nature in New York City* (Cambridge, MA: MIT Press, 2002), 122.

72 Ibid., 125; J. Vogel, "Opening Ways From the City to North, East, South, West," *New York Times*, January 10, 1932, A24; Williams, *City of Ambition*, 184.

73 Cutler, *The Public Landscape of the New Deal*, 53.

74 Lewis Mumford and Robert Wojtowics, *Sidewalk Critic: Lewis Mumford's Writings on New York* (Princeton Architectural Press, 1998), 225.

75 Caro, *The Power Broker*, 559.

76 Williams, *City of Ambition*, 196.

77 Caro, *The Power Broker*, 533.

78 Ibid., 557–8.

79 David Milton, *The Politics of U.S. Labor: From the Great Depression to the New Deal* (New York: Monthly Review Press, 1982), 73–4.

80 Ibid.

81 Irwin M. Marcus, "A Century of Struggle in Homestead: Working-Class Responses to Corporate Power," in *U.S. Labor in the Twentieth Century: Studies in Working-Class Struggles and Insurgency*, ed. Paul Le Blanc and John H. Hinshaw (Amherst: Humanity Books, 2000), 344.

82 Roy Lubove, *001: Twentieth Century Pittsburgh Volume 1: Government, Business, and Environmental Change* (Pittsburgh: University of Pittsburgh Press, 1996), 5.

83 Hinshaw, *Steel and Steelworkers*, 42.

84 Milton, *The Politics of U.S. Labor*, 31.

85 Hinshaw, *Steel and Steelworkers*, 63; Paul Le Blanc and John H. Hinshaw, eds., *U.S. Labor in the Twentieth Century: Studies in Working-Class Struggles and Insurgency* (Amherst: Humanity Books, 2000), 346.

86 Milton, *The Politics of U.S. Labor*, 105.

87 Hinshaw, *Steel and Steelworkers*, 3.

88 Warren, *Big Steel*, 167.

89 Le Blanc and Hinshaw, *U.S. Labor in the Twentieth Century*, 346.

90 Milton, *The Politics of U.S. Labor*, 104.

91 Hinshaw, *Steel and Steelworkers*, 4–5.

92 Sherie R. Mershon and Joel A. Tarr, "Strategies for Clean Air: The Pittsburgh and Allegheny County Smoke Control Movements, 1940–1960," in *Devastation and Renewal*, 148.

93 David Harvey, *A Brief History of Neoliberalism* (Oxford: Oxford University Press, 2007), 44–5.

94 Thomas J. Main, *Homelessness in New York City: Policymaking from Koch to de Blasio* (New York: NYU Press, 2017), 14–15.

95 Mark Singer, *Dark Days*, Film (New York: Dogwoof Productions, 2000); Margaret Morton, *The Tunnel: The Underground Homeless of New York City – The Architecture of Despair* (New Haven: Yale University Press, 1995).

96 Morton, *The Tunnel: The Underground Homeless of New York City*, 25.

97 Ibid., 22.

98 Rogers, *An Economic History of the American Steel Industry*, 127.

99 Ibid., 149–50.

100 Willard E. Miller, *A Geography of Pennsylvania* (University Park: Pennsylvania State University Press, 1994), 291.

101 Hinshaw, *Steel and Steelworkers*, 244.

102 Ibid., xiii.

103 Mark Roth, "Homestead Works: Steel Lives in Its Stories," *Pittsburgh Post-Gazette, post-gazette.com*, July 30, 2006, http://old.post-gazette.com/pg/06211/709449-85.stm.

104 R. Timothy Sieber, "Waterfront Revitalization in Postindustrial Port Cities of North America," *City & Society* 5, no. 2 (1991): 120–36.

105 Niall Kirkwood, ed., *Manufactured Sites: Rethinking the Post-Industrial Landscape* (London and New York: Taylor & Francis, 2001).

106 Elizabeth Meyer, "Uncertain Parks: Disturbed Sites, Citizens, and Risk," in *Large Parks*, eds. Julia Czerniak and George Hargreaves (New York: Princeton Architectural Press, 2007), 58–85.

107 Redevelopment Authority of Allegheny County, "Carrie Furnace Redevelopment Site Request for Development Proposal, Revision 1," accessed December 15, 2018, www.alleghenycounty.us/economic-development/bids/carrie-furnace-site.aspx.

108 Erin Deasy, Allegheny County Economic Development, Personal communication, July 20, 2016.

109 Rick Darke, "Addition by Reduction Project at Carrie Furnaces," *Rivers of Steel Arts*, http://rosarts.org/programs/eco-arts.

Bibliography

Baker, Orrin Hugh. "The Development of the Steel Industry in America." Thesis for B.Sc. in Mechanical Engineering, University of Illinois, 1907.

Baraff, Ron. Personal communication, July 10, 2014.

Bostaph, Samuel. *Andrew Carnegie: An Economic Biography*. Lanham: Rowman & Littlefield, 2017.

Carnegie-Illinois Steel Corporation. *Growing with America*. Pittsburgh: The Corporation, 1948.

Caro, Robert A. *The Power Broker: Robert Moses and the Fall of New York*. New York: Vintage, 1975.

"Contracts for Parkway Bridge Let 4 Hours After City Approves Plan." *New York Times*, June 15, 1935.

Cutler, Phoebe. *The Public Landscape of the New Deal*. New Haven: Yale University Press, 1985.

Darke, Rick. "Addition by Reduction Project at Carrie Furnaces." *Rivers of Steel Arts*. Accessed January 21, 2019, http://rosarts.org/programs/eco-arts.

David Harvey. *A Brief History of Neoliberalism*. Oxford: Oxford University Press, 2007.

Deasy, Erin. "Allegheny County Economic Development." *Personal Communication with Author*, July 20, 2016.

Duffus, Robert L. "Era of New Splendor Opens for West Side." *New York Times*, July 14, 1929.

"Extending West Side Highway to the North." *New York Times*, January 5, 1936.

Gandy, Matthew. *Concrete and Clay: Reworking Nature in New York City*. Cambridge, MA: MIT Press, 2002.

Goetzmann, William, N., and Frank Newman. "Securitization in the 1920's." *NBER Working Paper Series*. Cambridge, MA: National Bureau of Economic Research, 2010, https://doi.org/10.3386/w15650.

Grimm, Edward. *Riverside Park: The Splendid Sliver*. New York: Columbia University Press, 2007.

Henshaw, Robert E. *Environmental History of the Hudson River: Human Uses that Changed the Ecology, Ecology that Changed Human Uses*. Albany: State University of New York Press, 2011.

Hinshaw, John, H. *Steel and Steelworkers: Race and Class Struggle in Twentieth-Century Pittsburgh*. Albany: State University of New York Press, 2002.

Historic American Engineering Record. *Henry Hudson Parkway*. New York: Historic American Engineering Record, 2006, https://riverdalenature.org/wp-content/uploads/2013/06/Historic-American-Engineering-Record-HAER-report-on-the-Henry-Hudson-Parkway.pdf.

"Iron Floats to Market." *Saturday Evening Post*, December 23, 1939.

Jennings, Edward P. "The Mesabi Iron Range." *Science* 23, no. 575 (February 1894): 73.

Jensen, Jens. *Report to the League for the Protection of Riverside Park, November 29, 1916*. New York: Women's League for the Protection of Riverside Park: 1916.

Joel A. Tarr, and Terry F. Yosie. "Critical Decisions in Pittsburgh Water and Wastewater Treatment." In *Devastation and Renewal: An Environmental History of Pittsburgh and Its Region*, edited by Joel A. Tarr, 64–88. Pittsburgh: University of Pittsburgh Press, 2003.

Kirkwood, Niall, ed. *Manufactured Sites: Rethinking the Post-Industrial Landscape*. London and New York: Taylor & Francis, 2001.

Le Blanc, Paul, and John H. Hinshaw, eds. *U.S. Labor in the Twentieth Century: Studies in Working-Class Struggles and Insurgency*. Amherst: Humanity Books, 2000.

Lubove, Roy. *001: Twentieth Century Pittsburgh Volume 1: Government, Business, and Environmental Change*. Pittsburgh: University of Pittsburgh Press, 1996.

Marcus, Irwin M. "A Century of Struggle in Homestead: Working-Class Responses to Corporate Power." In *U.S. Labor in the Twentieth Century: Studies in Working-Class Struggles and Insurgency*, edited by Paul Le Blanc and John H. Hinshaw. Amherst: Humanity Books, 2000.

McElwaine, Andrew S. "Slag in the Park." In *Devastation and Renewal: An Environmental History of Pittsburgh and its Region*, edited by Joel A. Tarr, 174–92. Pittsburgh: University of Pittsburgh Press, 2003.

Mencken, H.L. "The Libido for the Ugly." In *Prejudices: Sixth Series*, 187–93. New York: Alfred A. Knopf, 1927.

Mershon, Sherie R., and Joel A. Tarr. "Strategies for Clean Air: The Pittsburgh and Allegheny County Smoke Control Movements, 1940–1960." In *Devastation and Renewal: An Environmental History of Pittsburgh and its Region*, edited by Joel A. Tarr, 145–73. Pittsburgh: University of Pittsburgh Press, 2003.

Meyer, Elizabeth. "Uncertain Parks: Disturbed Sites, Citizens, and Risk." In *Large Parks*, edited by Julia Czerniak and George Hargreaves, 58–85. New York: Princeton Architectural Press, 2007.

Miller, Willard E. *A Geography of Pennsylvania*. University Park: Pennsylvania State University Press, 1994.

Milton, David. *The Politics of U.S. Labor: From the Great Depression to the New Deal*. New York: Monthly Review Press, 1982.

Morton, Margaret. *The Tunnel: The Underground Homeless of New York City – The Architecture of Despair*. New Haven: Yale University Press, 1995.

Muller, Edward K. "River City." In *Devastation and Renewal: An Environmental History of Pittsburgh and Its Region*, edited by Joel A. Tarr, 41–63. Pittsburgh: University of Pittsburgh Press, 2003.

Mumford, Lewis, and Robert Wojtowics. *Sidewalk Critic: Lewis Mumford's Writings on New York*. Princeton: Architectural Press, 1998.

New York Central Railroad. "Details of Frames 'Fla.b,' Roof Covering Over Tracks, W 98th St to W 111th St, New York City." August 1, 1936, Drawing 84780 (Olmsted Center, New York).

New York Central Railroad. "Framing Plans and Details, Express Highway & Track Covering, W. 94th St. to W. 98th St., New York City." February 1, 1936, Drawing 84063 (Olmsted Center, New York).

New York Department of Public Parks. *Third General Report of the Board of Commissioners of the Department of Public Parks*. New York: William C. Bryant & Co., 1875.

Olmsted, Frederick Law Jr. "Pittsburgh: Main Thoroughfares and the Down Town District." In *Report to the Pittsburgh Civic Commission*, 19–21. Pittsburgh: Committee on City Planning, 1910.

Redevelopment Authority of Allegheny County. "Carrie Furnace Redevelopment Site Request for Development Proposal, Revision 1." www.alleghenycounty.us/economic-development/bids/carrie-furnace-site.aspx.

Robbins, Leonard H. "Transforming the West Side." *New York Times*, June 3, 1934.

Rogers, Robert P. *An Economic History of the American Steel Industry*. New York: Routledge, 2009.

Roth, Mark. "Homestead Works: Steel Lives in Its Stories." *Pittsburgh Post-Gazette, post-gazette.com*, July 30, 2006, http://old.post-gazette.com/pg/06211/709449-85.stm.

Schopf, William J. *Cradle of Life: The Discovery of Earth's Earliest Fossils*. Princeton: Princeton University Press, 2001.

Sieber, Timothy R. "Waterfront Revitalization in Postindustrial Port Cities of North America." *City & Society* 5, no. 2 (1991): 120–36.

Singer, Mark. *Dark Days*, Film. New York: Dogwoof Productions, 2000.

Smith, Robert A.C. *The West Side Improvement and Its Relation to All of the Commerce of the Port of New York*. New York: M.B. Brown Printing & Binding Co., 1916.

"Smog Linked to Pneumonia: Blanket of Smoke and Fog Responsible for Death Rate in City of Pittsburgh." *The Globe and Mail*, November 6, 1937.

Stang, Ambrose H., Martin Greenspan, and William R. Osgood. "Strength of a Riveted Steel Rigid Frame Having Straight Flanges." U.S. Department of Commerce Research Paper RP1130. *Journal of Research of the National Bureau of Standards*, 21 (September 1938): 84780.

"Structural Steel Orders." *Wall Street Journal*, September 22, 1936.

Sweeny, Harry. *West Side Improvement: Published on the Occasion of the Opening October 12th, 1937*. New York: Moore Press, 1937.

Thomas J. Main. *Homelessness in New York City: Policymaking from Koch to de Blasio*. New York: NYU Press, 2017.

United States Department of the Interior, National Park Service. "Carrie Blast Furnaces Number 6 and 7." *National Historic Landmark Nomination*. Accessed February 13, 2018, www.nps.gov/nhl/find/statelists/pa/CarrieBlast.pdf.

Vogel, J. "Opening Ways From the City to North, East, South, West." *New York Times*, January 10, 1932.

Warren, Kenneth. *Big Steel: The First Century of the United States Steel Corporation 1901–2001*. Pittsburgh: University of Pittsburgh Press, 2001.

Warren, Kenneth. *The American Steel Industry 1850–1970: A Geographical Interpretation*. Pittsburgh: University of Pittsburgh Press, 1988.

Warren, Kenneth. *Wealth, Waste, and Alienation: Growth and Decline in the Connellsville Coke Industry*. Pittsburgh: University of Pittsburgh Press, 2001.

"West Side Project Ready to Hire 4,000." *New York Times*, May 28, 1934.

"West Side Tracks Plan is Complete." *New York Times*, April 8, 1916.

White, Langdon. "The Iron and Steel Industry of the Pittsburgh District." *Economic Geography* 4, no. 2 (April 1928).

Williams, Mason, B. *City of Ambition: FDR, LaGuardia, and the Making of Modern New York*. New York: WW Norton, 2014.

Willis, Carol. *Form Follows Finance: Skyscrapers and Skylines in New York and Chicago*. New York: Princeton Architectural Press, 1966.

Women's League for the Protection of Riverside Park. *Some Plain Facts About Riverside Park*. New York: Women's League for the Protection of Riverside Park, 1916.

Figure 4.1(A) Detail, juvenile London
plane trees and nursery worker, Rikers
Island Nursery

Source: New York City Parks Photo Archive.

Figure 4.1(B) Detail, planting London plane trees along Seventh Avenue, starting at 117th Street, 1959

Source: Photograph by William C. Eckenberg (The New York Times/Redux Pictures).

Chapter 4

Breathing with Trees

London Plane Trees from Rikers Island to Seventh Avenue, 1959

In a 1944 short story published in the literary journal *The Crisis: A Record of the Darker Races*, Florence McDowell narrates Lavinia, a high school student, strolling along tree-lined Seventh Avenue (renamed Adam Clayton Powell Boulevard in 1974), north of Central Park in Harlem:

> Surely there was more beauty on Seventh avenue than on any other street in Harlem. Where else could one feel the space that let her spread out inside? Where else could one see trees marching up the center? Her biology teacher had told her that they were Chinese plane trees. Lavinia liked the balls that hung like toy fruit or grace-notes after the leaves had given up, and she liked the rich brown and gray bark with its undergarment of cream and yellow. If she were walking on Morningside avenue, she could get close to such trees and sniff the bitter, teasing odor of the bark while it was still wet from the rain. Now the leaves were new with May. Those Seventh avenue plane trees did bring something to Harlem – more than could have been foreseen when they were planted.[1]

The trees that Lavinia admired were likely London planes (*Platanus x acerifolia*), the "x" indicating a cross between two species, one from Asia (Oriental plane, *P. orientalis*) and the other from the eastern United States (American sycamore, *P. occidentalis*). Born in a seventeenth-century botanical garden, where the two geographically distant trees might have been planted together, the vigorous London plane has since thrived.[2] Establishing itself in a rapidly urbanizing world, *P. x acerifolia* displayed resilience to the new atmospheric conditions of city life. In times of drought, the species sends out adventitious roots to secure moisture. As dust, heavy metals, and other particulates from the street gather on the tree's trunk, the London plane exfoliates bark in cloud-shaped pieces, revealing fresh gray-green surfaces below, effectively cleaning and protecting itself from contamination.[3] With a similar strategy, its maple-like leaves concentrate airborne particulate matter on their rough surfaces, shedding it with the rain.[4] Gathering, decomposing, and flushing particulates from the air to the soil and streets, the species improves its own health, but also the health of the people around it.

By the mid-century in New York City, the London plane was wildly popular as a street tree. As historian Thomas J. Campanella notes, the London plane

Figure 4.2 **London plane tree** (*Platanus x acerifolia*) **herbarium specimen, collected by M.H. Nee and D. Atha in the Bronx, 2011**

Source: Image courtesy of the C.V. Starr Virtual Herbarium of the New York Botanical Garden (http://sweetgum. nybg.org/science/vh/).

tree in New York City had a critical advocate in the young landscape architect Michael Rapuano. On his travels to Rome, Rapuano became enthralled with the masses of London plane trees that made up a third of Rome's shade canopy. Upon return, Rapuano worked for the Parks Department under Gilmore

Figure 4.3 **London plane trees in Riverside Park, New York City, 2015**

Figure 4.4 **New York City Department of Parks (and recreation) logo, inspired by the London plane Tree**

Source: New York City Parks Photo Archive.

Clarke on many New Deal–supported urban landscape projects (including the Riverside Park renovation discussed in Chapter 3), replicating the European *allées* and groves of robust London plane trees throughout the city.[5] The species had widespread appeal. It could survive where others couldn't. It actually "fit" in harsh urban conditions. Praising the species' ability to resist smoke, gas, foot traffic, and rough handling, a 1948 New York City Parks Department document proclaimed: "In short, it has the will to survive, and, unlike the arboreal primadonnas, it can take abuse."[6] Such admirable traits made the five-pointed leaf of *P. x acerifolia* the inspiration for the Parks Department logo, and sent its handsome silhouette multiplying throughout the boroughs.

Street trees are central characters in narratives about urban health. Often the only living element on a street, trees indicate a neighborhood's vitality.

While an abundant canopy of lush, robust, and long-living trees conjures associations of health, safety, and quality of civic life, a sparse, broken, dying canopy instead reflects disinvestment, social decay, and dis-ease. These associations are both ideological and real. On the one hand, trees in the city have long symbolized an implant of "nature"; they are seen as a natural salve – restorative and rehabilitative. On the other hand, as living organisms, trees die or thrive based on investment and maintenance; they register the systemic and spatial inequalities that exist within the city. When they flourish, they improve the air that people breathe in, the very stuff of human survival. As trees exhale, humans inhale. This chapter follows the mid-century wave of London plane trees from a nursery on Rikers Island to municipal planting projects throughout the city. I focus on a community-driven planting on Seventh Avenue in Harlem, north of Central Park, where demands for civil rights, equality, and basic community amenities in the final years of the first Jim Crow era extended to the street tree.

While taken for granted as part of the modern urban street, trees were not always considered in city planning. The Commissioner's Plan of 1811 projected a novel infrastructural logic for the city, but it was one without trees. The

AMERICAN ASSOCIATION

FOR THE

PLANTING AND PRESERVATION

OF

CITY TREES

HEADQUARTERS: CHILDREN'S MUSEUM IN BEDFORD PARK
BROOKLYN AVE. AND PARK PLACE, BROOKLYN, N. Y.

Improve Your Street: Plant Trees
Get Your Neighbors to Cooperate

BY J. J. LEVISON, M. F.
FORESTER

ORIENTAL SYCAMORE TREES, SO. PORTLAND AVE., BROOKLYN, N. Y.
PLANTED BY COOPERATIVE SYSTEM. PHOTO TWO MONTHS AFTER PLANTING

Figure 4.5 Brochure cover, J.J. Levison, *Improve Your Street: Plant Trees Get Your Neighbors to Cooperate*, Brooklyn: American Association for the Planting and Preservation of City Trees, 1912

existing urban shade trees were removed as the street grid was extended through the city.[7] Over the nineteenth century the city invested in a complex assembly of pavement, drainage, sidewalks, curbs, and lighting, but the planting of trees along those roads received little attention.[8] Instead of being included in the comprehensive, modernizing grid, trees were planted at the whim of property owners, without official oversight regarding aesthetics, health, or maintenance. Into the early twentieth century, advocates encouraged homeowners to coordinate with their neighbors to beautify their streets: "If you are prompted by civic pride to become the moving spirit in the realization of your street beautiful," one brochure advised, you should determine how many trees on a block were needed, coax your neighbors to share the cost, and then procure a planter and the specimens.[9] The result was a random composition of ill-fated trees.

Critics mocked the "ghastly parodies" of trees planted in concrete.[10] They were seen as vulnerable outsiders from the country, poorly suited to the urban landscape and with little chance of survival. Despite their well-known success in cities like Washington, DC and Paris, naysayers argued that they couldn't survive the hardships of impermeable surfaces, hot and reflective materials, horse bites, rambunctious kids, oil, salt, and air pollution. As a 1902 *New York Times* editorial argued, "It is better to limit tree planting to one tree a year and have that a possible ornament and benefit to the city than to plant a million and have an equal number of failures."[11] Furthermore, these sickly trees compromised and concealed architecture's clean forms with their sad and messy bodies. Trees might fit in the domestic sphere, but not the commercial one. New zoning regulations had separated commercial and industrial from residential zones, enabling those opposed to street trees to exclude them in the ever-expanding commercial zones of the city.[12] From this perspective, trees belonged enclosed in parks – encircled models of nature – not on streets. But others saw the street tree as an antidote to urbanization: a natural salve to a harsh environment. This perspective, however, still idealized the tree as an object of a purely external nature.

In the early twentieth century, street trees would develop a new character other than outsider or benevolent natural transplant: they became the charismatic flora of the progressive health movement, and the new "tree-conscious citizen" was their agent.[13] Tenement districts suffered from incredible heat in summer months, as buildings and pavements soaked up the sun with little shade and no means to circulate air. As disturbingly high death rates were linked with urban summer heat, public health advocates rallied for more trees.[14] Doctors, landscape architects, and citizen activists convened the Tree Planting Association (TPA) in 1897 to call for the adoption of scientific tree culture throughout the city's boroughs. Leaders such as Dr. Stephen Smith, who was deeply invested in the health reform movement, championed the street tree as a remedy to the toxic urban environment. The citizens' very breath, Smith underscored, was enmeshed with the tree's existence:

> As we enter the shade of a tree in full leafage, on a hot Summer day, we feel a thrill of energy which quickens our footsteps, expands our chests,

brightens our thoughts, and gives a new impulse to all our vital processes. What has happened? We have thrown out of our lungs the depressing dioxide and replaced it with exhilarating oxygen from the nearby tree.[15]

The Association argued for a scientific understanding of urban tree function. One commissioned study found that a single elm tree transpired 260 barrels of water per day through its 5 acres of foliage. Dr. Smith interpreted these findings: "From this investigation we learn that a tree standing by our dwelling in the city and lifting its foliage in the air, story by story, would bring to every window which it passed acres of park scenery. In the hot Summer days and nights it would purify the air entering the chamber and cool it with a delicious moisture."[16] Dr. Smith describes street trees at once as a vertical park, a heat- and moisture-regulating companion of the building stock, and a vital partner in health for New York's citizens. The street tree, Smith and the TPA argued, should become an integral part of the modernizing street assembly, offering it a new biological function.

No longer just a tree, the *street tree* became a unique horticultural specimen. Unlike garden and park trees, street trees needed to be hard, straight, immune from insect attacks, shady, clean, and long-living. And the city would need to acquire them from a nursery where they would be properly trained to survive transplantation. Commercial nurseries on Long Island and in New Jersey, well established for garden plants and ornamentals, subsequently developed expertise in street tree production. Ideally, one report advised, urban street trees should come from municipally-owned nurseries, where city workers could trim taproots and encourage the lateral root systems that would keep trees stable in shallow urban soil, trim the lower branches to allow for a taller canopy that could suspend over the traffic, and prune trees to give them strong single leaders.[17] A municipal nursery would offer an ongoing stock of street trees, grown to the correct specifications of a uniform planting system; it would also be cheaper, and anticipate long-term needs.

The TPA called for a Tree Culture Bureau within the Parks Department, staffed with foresters, to coordinate and maintain the plantings. Chapter 453 of the New York state legislature had given the Parks Department jurisdiction over all vegetation in streets and public spaces as early as 1902, but there was no allotted funding to pay for trees for several decades.[18] With a limited budget, the Parks Department oversaw the planting and maintenance of several decades of street trees that continued to be paid for by individuals. The Tree Planting Association envisioned a "street tree system" – a legible network of planted avenues connecting the otherwise scattered existing parks. Such a network of "inter-park connections," landscape architect Laurie Davidson Cox, argued, would create a unique species of park system, consistent with New York officials' desire to create one of the world's most beautiful cities.[19] While Cox's study found that many of Manhattan's streets could not support trees, he recommended that planting be focused in residential areas. With a network of orthogonal red lines and polygons, Cox mapped a nearly continuous system of streets that could connect significant parks. Blue lines indicated streets that, by

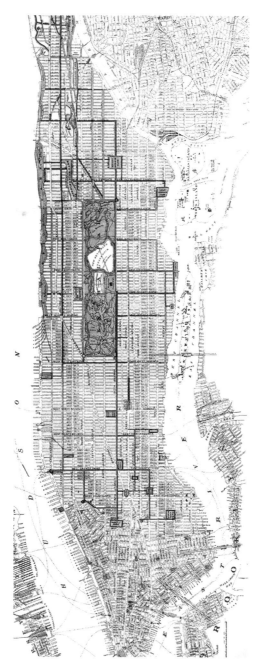

Figure 4.6 Detail, Laurie Cox's "Proposed System of Tree
Planting for the Borough of New York" included streets
that already had satisfactory tree growth, including Seventh
Avenue north of Central Park, 1916

Source: Laurie Davidson Cox, A Street Tree System for New York
City, Borough of Manhattan. Syracuse: Syracuse University, 1916.

1916, had already had satisfactory tree plantings. At the time, Seventh Avenue in Central Harlem north of Central Park was among these few major boulevards already lined with a respectable canopy of trees.

Indicator Species

By the late nineteenth century, Seventh Avenue north of Central Park was envisioned as a grand, leafy boulevard for upscale living.[20] Once a remote ex-urban escape for the wealthy, the area became a primarily mixed working-class German and Irish neighborhood, gaining Jewish and Italian immigrants, and some African-American residents towards the turn of the century.[21] When plans for a new subway line under Lenox Avenue were announced in the 1890s, developers built luxury apartment buildings centered on a newly widened and planted Seventh Avenue.[22] Flanking the avenue, towering American elm trees shaded broad sidewalks; extending down the middle, a generous median strip was planted with grass and punctuated with stately lighting standards. Developers advertised Seventh Avenue as a boulevard for new park living, and architects designed models of new luxury apartment buildings along the tree-lined avenue. Architects Clinton & Russell's 1901 massive Graham Court (commissioned by real-estate tycoon William Waldorf Astoria), for example, referenced an Italian Renaissance palazzo with its massive central courtyard and elliptical

Figure 4.7 **Trees on Seventh Avenue, looking south from 121st Street**
Source: Photograph by Thaddeus Wilkerson (Museum of the City of New York, F2011.33.1577).

garden, demonstrating new models for genteel, higher-density living complete with modern conveniences and powered by elevator technology.[23] This modern living was restricted, however; Graham Court and many other apartment buildings didn't admit black residents until the late 1920s.[24]

In Harlem – like elsewhere in the city – the speculative construction boom produced a real-estate bubble, as well as numerous vacant high-cost rental units. The bubble burst in 1904, driving Harlem landlords to adjust racist restrictions to recoup their losses. Owners of white-only buildings started renting to black tenants, regularly charging them twice that of whites for the same unit.[25] Despite this discrimination, black New Yorkers began moving uptown, displaced as historically black neighborhoods in Wall Street, the Tenderloin District, and Seneca Village became the city's financial center, Penn Station, and Central Park, respectively.[26] At the same time, large numbers of southern African Americans, seeking industrial jobs and fleeing violence from the atrocities of Jim Crow laws, began the Great Migration to northern cities, including New York; those arriving gravitated to Harlem.[27] The 1900 Tenderloin District race riots further motivated black New Yorkers to move north to Harlem's growing black communities.[28] Finally, waves of immigrants from the Caribbean chose Harlem as their preferred neighborhood to land. By the 1930s, 200,000 African Americans lived in Harlem, gaining access to unusually decent housing (although at discriminatorily high rents) alongside gracious tree-lined boulevards.[29]

"Negro Harlem is best represented by Seventh Avenue," Harlem Renaissance writer Thurman Wallace wrote in 1928:

> It is a grand thoroughfare into which every element of Harlem population ventures either for reasons of pleasure or business. . . . Seventh Avenue is majestic yet warm, and it reflects both the sordid chaos and the rhythmic splendor of Harlem. From five o'clock in the evening until way past midnight, Seventh Avenue is one electric-lit line of brilliance and activity. . . . Dwelling homes are close, overcrowded and dark. Seventh Avenue is the place to seek relief. People Everywhere . . . A medicine doctor ballyhooing, a corn doctor, a blind musician, serious people, gay people, philanderers and preachers. Seventh Avenue is filled with deep rhythmic laughter. It is a civilized lane with primitive traits, Harlem's most representative street.[30]

In his account of the Harlem Renaissance and his own youth there, basketball legend Kareem Abdul-Jabbar describes Seventh Avenue, lush with trees and flowers on the median strip, as a "Garden of Eden."[31] Along this garden avenue stood many of the great theaters and jazz clubs, hotels, and African and African American–focused bookstores, churches, and social institutions. Street life intensified around church entrances, theater box offices, and significant cultural landmarks. One of these was a street tree. "The Wishing Tree" (later known as "The Tree of Hope"), an elm located between the famous Lafayette Theatre and Connie's Inn, was said to give luck to musicians about to perform, or grant wishes to those who rubbed its bark.[32] The tree was cut down in 1934 when

Figure 4.8 **Wishing Tree, Seventh Avenue at 132nd Street, 1936**
Source: From the New York Public Library.

Figure 4.9 **Seventh Ave looking uptown from 125th Street, 1945**
Source: Photograph by Nichols (Photographs and Prints Division, Schomburg Center for Research in Black Culture, The New York Public Library).

a street widening took down the street's sidewalk canopy. While the culture of African-American Harlem was soon recognized as a phenomenon around the world, racial inequities within the culture industries were blatant: many clubs starring black musicians catered to white patrons only; whites owned the clubs, owned the record companies, and profited from Harlem's cultural capital. As poet Langston Hughes put it: "Downtown: *white*. Uptown: *black*. White downtown pulling all the strings in Harlem."[33]

Seventh Avenue was a significant space for community representation and political demonstration. As James Weldon Johnson wrote in 1930, Harlem is also a parade ground. Almost any excuse for parading is sufficient – the funeral of a member of the lodge, the laying of a corner stone, the annual sermon to the order, or just a general desire to "turn out."[34] Parades by Marcus Garvey and the Universal Negro Improvement Association were widely documented and famously elaborate; brass bands, decorated men and women from the Africa Legion, Black Cross Nurses, and others bearing pan-Africanist slogans paraded up Lenox Avenue and down Seventh. Parades brought local pride, political messages, and community solidarity to the street itself, registering the intense amount of organizing happening throughout the neighborhood.[35] Historian Shannon King argues that Depression-era Harlem was a "training ground for freedom fighters" as residents organized to fight the systematic racial violence of everyday life – police brutality, and obstacles to housing, jobs, and economic empowerment – and discrimination at every turn. Grassroots organization helped build the social infrastructure, institutions, and momentum that fueled resistance in the decades to come.[36]

As Harlem's population shifted from predominantly white to black, white landowners devised new mechanisms for exclusion and exploitation. In addition to charging discriminatory rents, landlords practiced block-busting, and white tenant associations produced covenants prohibiting owners from renting or selling to black tenants and black businesses, or from hiring more than a certain number of black employees.[37] The 1934 National Housing Act redlined black neighborhoods, limiting mortgage availability and preventing ownership for those who would have been able to buy.[38] Construction of new housing stopped, just as more and more people poured into the neighborhood in need of it, and residents were forced to crowd into apartments. In short, private landowners uniformly disinvested in the maintenance of building stock and amenities while continuing to profit from high rents and high demand from a lower-waged population.

This disinvestment manifested in the physical street itself – in decaying housing stock, improper garbage management, rat infestations, and health epidemics. Harlem's housing conditions were deemed "deplorable," and the neighborhood was considered the second-most "disease-ridden" in the city.[39] Following World War II, as New York City lost manufacturing jobs to low-wage and non-unionized jobs, Harlem's population suffered disproportionately. In McDowell's story from the beginning of this chapter, written at the height of World War II, Lavinia experiences the physical confluence of these struggles. While the plane trees on Seventh Avenue give her a sense of openness, the rest of her world is compressed. The journey to her cramped apartment is full of obstacles: piled-up garbage, relentless noise, and foul smells. As a young poet, Lavinia struggles and succeeds to articulate beauty within the contradictions and adverse physical conditions of her neighborhood.

Building on strategies and social infrastructures from the early century, Harlem activists fought the full spectrum of disinvestments and discriminations that impacted the community's physical environment and health. One group, the People's Civic and Welfare Association (PCWA), linked the struggle for basic civil rights with the right to a clean, safe, and healthy street. In addition to calling out atrocious housing conditions,[40] summonsing negligent landlords for the piles of garbage in the streets,[41] and citing the Department of Sanitation for incinerators in violation of the Department's own Sanitary Code,[42] they focused much of their activity on cleaning up and improving the quality of the streets. Seventh Avenue, they argued, was a grand and historic boulevard and deserved the same treatment as Park Avenue or Fifth Avenue downtown.[43]

Starting in 1946, the PCWA petitioned Robert Moses and the Department of Parks for a "civic beautification plan" of Seventh Avenue from 110th to 145th Street.[44] They demanded tree planting along the street's sidewalks (where street-widening measures in the 1940s had removed the rows of American elm trees, including the Wishing Tree), traffic calming measures, appropriate lighting, and street and curb repair. They argued for shrub planting down the median strip, surrounded by protective fences, similar to Park Avenue. In a plea for community support for trees, "Plain Talk" columnist Elmer A. Carter wrote,

> Can you imagine what Harlem would look like without any trees? . . . We need their shade and we need the refreshing breezes in the heat of the noonday sun. And we need their beauty to keep us from thinking that all the world is hard and is made of brick and stone.[45]

The PCWA's Glester Hinds believed that the quality of the street would be a form of "visual education" and a source of morale for residents who had grown accustomed to the declining landscape.[46] Within the broader struggle for civil rights, the PCWA asserted the right to a healthy and beautiful physical environment.

Hulan E. Jack, the Harlem State Assembly Representative-turned-Manhattan Borough President, also recognized the power of a clean, planted street to promote civic health. Jack's election in 1953 as the first African American to serve as Borough President (alongside Harlem-born Congressman Adam Clayton Powell Jr., who was elected just years before) signaled a landmark of political representation of – and advocacy for – Harlem. Within a political platform focused on combatting racial discrimination, Jack positioned tree planting as a catalyst for local morale and civic participation. Imagining the grand planted boulevards of Paris, Jack pronounced, "our intention is to make Manhattan as pleasant a place to live in as it used to be in our grandfathers' day. . . . We want

Figure 4.10 **Planting London plane trees along Seventh Avenue, starting at 117th Street, 1959**

Source: Photograph by William C. Eckenberg (The New York Times/Redux Pictures).

to lure all those poor exiles in the suburbs back to town, where we're sure that, in their hearts, they still long to be."[47] These "exiles" were droves of primarily white, affluent residents fleeing Manhattan – increasingly associated with urban decline, racist fears, and crime – for the suburbs. The borough's trees were failing and dying, and Jack recognized their symbolic and aesthetic significance; their death signaled the feared urban decline, while new plantings signaled investment and community care. Jack promoted mass tree plantings as a means to reestablish civic pride and signal neighborhood investment, and followed the PCWA's suggestions for Seventh Avenue.

On a spring day in 1959, in response to the PCWA's thirteen-year-long request, Hulan E. Jack ceremonially shoveled soil on the base of a newly positioned London plane tree at Seventh and 117th Street. This was the first of 398 trees slated to line the sidewalks from 110th to 155th, planted every 40 to 50 feet. To complete the $600,000 plan, the city's first fluorescent lights, twice as bright as those existing,[48] would be installed down the center mall parallel to the trees. Large street signs would replace old ones, and broken curbs and streets would be repaired.[49] Jack hoped that the improvements of Seventh Avenue would motivate adjacent property owners to improve and renovate their dilapidated buildings, and return the avenue to its former grandeur.[50] After the ceremonial planting, Jack asked residents to take care of the young trees: "Someday," he remarked, "they will be thirty feet tall."[51]

When the London plane trees arrived, they were just 8 feet tall. The juvenile trees had grown up on Rikers Island, just a few miles away in the East River, almost at Flushing Bay. Before they lined the sidewalks, they had stood in nursery rows oriented east–west like runways that took off into the Bay. For seven years they had been cared for – planted, transplanted, watered, fertilized, pruned, monitored – by men incarcerated in the Rikers Island penitentiary complex. They had formed straight and sturdy trunks, an optimal 3 inches in diameter, a sturdy single leader, and a carefully pruned branching structure. They had fed from the nutrients of a former dump, with currents, barges, and airplanes passing by. They faced the northern wind, thickening their wood in its direction.

Rikers Terroir

Since 1884, Rikers Island has been an uncanny landscape of incarceration, disposal, and cultivation in a direct feedback loop with other places in New York City. The New York Commission of Public Charities and Correction purchased the originally 42-acre island from the Rycken family as a work-house and penal colony. There, prisoners grew vegetables and reared pigs (with scraps from the city's restaurants), returning thousands of pounds of produce and pork to the city's institutions – at no cost.[52] Over time it became a fully operational landfill (later under the control of the Department of Sanitation), initiating the miserable concept of mixing incarceration with a dump. With a constant stream of garbage and

Figure 4.11 **Aerial view of Queens with Rikers Island at left, 1931**

Source: Photograph by Fairchild Aerial Surveys, Inc. (courtesy New York City Municipal Archives).

Figure 4.12 **Aerial view of Rikers Island, circa 1935**

Source: Photograph by McLaughlin Aerial Surveys (courtesy New York City Municipal Archives).

subway excavate, 42 acres grew to 600. Loads of garbage arrived on the island by scow and then traveled up inclined rail lines to the dump's peak. On the southern portion of the island, garbage hills reached 100 feet high,[53] and over time the material inputs changed from homogenous, minimal, and organic, to heterogeneous, voluminous, and toxic.

By the late 1930s, the island was a "mountainous" garbage dump, one journalist recalled, "on which rats disported like Alpine goats and where 'eternal' fires smoldered like volcanoes."[54] As the World's Fair of 1939 approached, the dump – which could be seen and smelled from the Flushing Meadows fair site – threatened to embarrass the city and was shut down. In its place, a $10-million "model" penal laboratory was planned. There, nine buildings, with modern architectural details (such as automatic lockdown of all cells) would form a complex that looked less like a penitentiary and "more that of an expensive hotel." Drawing from the emerging science of human behavior, the Department of Corrections claimed that they would assess inmates' physical and mental conditions, and in some cases the environment and family history of incoming prisoners, and work towards reform, not simply discipline.[55] As plans for the penitentiary developed, so did plans to increase the productive potential of the former dump landscape.

Addressing delegates of the Fifteenth National Shade Tree Conference, Mayor La Guardia announced a plan for a city nursery on Rikers Island. The nursery would be directed by Park Commissioner Robert Moses, staffed by the Parks Department, and tended entirely by prisoners.[56] It would produce an endless supply of affordable shade trees for the city's parks and streets, grown to the city's various specifications. La Guardia assured the delegates that, while New York's trees had long been mistreated and disregarded, a new movement was taking hold. The plan was a resounding win-win, as one 1941 press release announced:

> The development and maintenance of a tree nursery on Riker's Island, using penitentiary labor, is of immense benefit to the city. In addition to replacing an unsightly and odorous nuisance and providing healthful occupation to the prison inmates the nursery will soon produce, at relatively small cost, an annual crop of shade and ornamental trees for parks, parkways and park streets.[57]

The nursery project was a story of redemption. A former dump would be rehabilitated through the continual growth of thousands of trees, which would then transform New York City from its perception as a "vast treeless city of stone" into one "dotted with shade trees."[58]

Rikers offered ample nursery grounds in a location close to the boroughs of New York City, yet undesirable for other uses. For decades, sourcing enough trees for park and street planting projects had been a major challenge. Even with a well-established private nursery industry nearby, and with multiple city-run nurseries in city parks, the Parks Department could often not procure enough trees. Furthermore, land pressure for recreational spaces was so great that park-based city nurseries were difficult to justify, and labor costs were high. On Rikers Island, however, there was no recreation pressure, and a strong motivation to reclaim the dump and exploit the labor of its incarcerated residents.[59] Free land and an incarcerated workforce meant that Rikers trees would cost just $30 each to plant, compared to the $60 private nursery tree.[60] Anticipating 10,000 replacement trees a year, the Rikers nursery was a benefit to the budget.[61]

"It's good, healthy work for them and an ideal form of outdoor exercise for the fellows," Robert Moses explained.[62] Not only did the job provide a chance to do constructive work outside, it entailed caring for and rearing living plants. Such a hearty and nurturing vocation would help improve the inmates while they produced valuable stock. Politicians and journalists emphasized how the street trees from Rikers Island – unlike the sickly, ill-fitting trees of the past – would be grown on New York City's very waste, and reared by its street-hardened criminals. The inmates came "from the city pavements," the tough urban conditions that the street trees they cared for would be uniquely suited to.[63] Rikers Island soil and labor would form the *terroir* – the particular biophysical and cultural conditions that ensured the success or fitness of the land's fruits. Because of these conditions, one journalist speculated, "The Rikers Island babies, set out in the city's parks, stand a better chance for survival than most New York trees."[64] Unlike your typical juvenile tree, these ones would survive because of where they came from, and they would truly belong in New York City.

Two hundred Rikers inmates leveled peaks of garbage and ash from the southeastern limits of the island. They deposited soil, and combed through the top 2 feet to remove bedsprings, tires, bottles, and scrap metal – anything too large or inert to compost. They transplanted a massive swath of silver poplars, recovered from Orchard Beach, as a broad, protective windbreak to the island's east and south sides. They installed snow fences around the perimeter. They planted cover crops – perennial rye and hairy vetch in the winter, and millet and cow peas in the late summer – and turned them under to cultivate and improve the soil. They dug miles of trenches to receive the incoming seedlings. In the newly constructed nursery grounds, a heterogeneous mixture of fermenting matter steamed constantly. Escaping methane gas and mystery "hot spots" reminded planters of the dump below the soil, and produced a randomly distributed pattern of mortality in the young trees.[65] With each successive cycle of cover crop, and the continual removal of garbage from the earth, the soil texture improved; in fact, according to Chief Horticulturalist David Schweizer, New York's compacted wastes made "amazingly fertile soil."[66]

Just two years after marking out the new nursery, almost 13,500 saplings, primarily London plane, American elm, and Norway maple (but also pin oaks, silver lindens, ginkgoes, and honey locusts) had already extended their roots into miles of prepared trenches,[67] and 15,000 more were on their way. The young plane trees growing on Rikers, descendants of specimens from Bloodgood's Nursery in Albertson, Long Island, multiplied through propagation.[68] In

Figure 4.13 **Rikers Island, circa 1925**

Source: Photograph by Percy Loomis Sperr (© Milstein Division, The New York Public Library).

Figure 4.14 **Rikers Island Nursery in front of prison building, 1941**

Source: New York City Parks Photo Archive.

Figure 4.15 **Rikers Island land being prepared for sowing of alfalfa and rye, 1937**
Source: Courtesy New York City Municipal Archives.

the depths of winter, workers on Rikers cut 12-inch-long segments, each with four buds, from the lower branches of young plane trees. This removed some of the undesirable lower branching from the trees, while providing wood cuttings for propagation. They bundled cuttings into groups of fifty, placing them into pits of sand and salt hay, with the eyes down. Come spring, they planted the cuttings into beds to root for two years before transplanting them again to tightly spaced nursery rows, and then, as the saplings took shape, transplanting them again to give them more room to grow. For the first two years in the nursery row, workers would allow the plane trees to sprout shoots, but afterwards they would trim these, carefully monitoring the trees to balance desired root growth, height, and caliper size of the trunk.[69]

While the trees were in the nursery rows for a few years, workers continued to water, spray, and prune – shaping the trees to develop a strong central leader, and a high branching structure that would later extend above and over sidewalks. By the time that a London plane tree was dug out of the ground, its root ball trimmed and wrapped in burlap for shipment, it had received multiple cycles of seasonal maintenance and care. The famously fast-growing London planes grew the requisite 2.5- to 3-inch diameter trunk, and an 8- to15-foot

Figure 4.16 **Seedlings, Rikers Island tree nursery**
Source: New York City Parks Photo Archive.

Figure 4.17 **Plane tree saplings, Rikers Island tree nursery, 1941**
Source: New York City Parks Photo Archive.

Figure 4.18 **Juvenile plane trees and nursery worker, Rikers Island Nursery**
Source: New York City Parks Photo Archive.

tall canopy within six or seven years. Though Rikers inmates served sentences of three years or less, with their stints at the tree nursery even shorter, they still would have seen some visible change in their charges, from seedlings to wispy saplings to sturdy adolescents, before being shipped away. The trees left Rikers by ferry until a bridge to Queens was constructed, and then they left by truck.

Two years after the first planting, the trees were growing faster and stronger than anyone could have predicted. The first generation of trees was ready for transplant a couple of years earlier than expected, and the Parks Department quickly made plans to extend the nursery to include 50,000 more. Given the nursery's demonstrated success, the dream of a steady supply of 10,000 trees per year was realized; these would not just replace the 7,000 trees dying each year to age, storms, and vandals – they would proliferate throughout the city.[70] With anticipation, the first delivery of Rikers trees – 146 six-year-old planes – were installed in the parking area on Beach Channel Drive, Rockaway, east of the Cross Bay Veterans Memorial Bridge. "The trees are smaller than we usually plant," an assistant to Robert Moses reported in a memo, "but they are well shaped."[71] The nursery grew from 20 to 115 acres,[72] filling out the eastern part of Riker's Island, just 100 feet from the newly constructed runways of LaGuardia Airport. And the press was quite enamored with the new project, with one reporter writing: "With 100,000 trees and shrubs growing on Rikers

Island our five boroughs have an arboreal backlog which promises a green future for many a present asphalt desert."[73]

Rikers' previous activities and natural history had produced a unique biophysical environment – a landfill substrate, an island condition free of predators, a sparse human population, jet fuel and dust from LaGuardia Airport just hundreds of feet away, and thousands of differently aged trees rooted in the soil. Over time, these conditions would come to support a range of bird species, including pheasants, cormorants, red-winged blackbirds, red-tailed hawks, ducks, and geese.[74] "Looks like a bird sanctuary here now, doesn't it?" David Schweizer said as he led a *New Yorker* reporter around the nursery.[75] While rat populations, infamous at the time of the dump's closure, had quelled, they still roamed freely, sometimes gnawing on trees and creating headaches for the Parks Department.[76] A relatively quiet operation, nursery life was punctuated with infrequent drama: an unexplained infestation of black widow spiders in 1944; a 1955 hurricane that flooded the fields with salt water, wiping out 4,000 London plane trees;[77] a Northeast Airlines plane crash into the nursery, which plowed through 666 London planes trees in 1957,[78] killing 20 people onboard. And in 1966, two prisoners temporarily escaped capture by hiding in a foxhole between rows of London planes for days, only to be discovered when they went looking for water.[79]

"Sure, it's prison labor," Rikers Warden Anna M. Kross explained in 1964. "The Department of Parks gives us the orders and we supply the labor." While the size of the nursery operation fluctuated over the years, at its peak up to 5 Parks Department employees supervised as many as 290 prisoners (who were in turn accompanied by 9 corrections officers[80] and worked a morning and afternoon shift each day).[81] Prison work programs were promoted on Rikers to rehabilitate and "correct" inmates, but equally to exploit a large, captive, and cheap workforce for profit. The rehabilitative aspect made for more positive journalism. Since the pre-Rikers prison colony operated its municipal farm and piggery, the island had always generated income through incarcerated labor. In addition to tending trees, inmates baked for the city schools, repaired mattresses for the Department of Welfare, and fixed vehicles and sewed uniforms for the Department of Corrections. Beyond the opportunity for inmates to train for future employment, gain discipline, and build social skills, nursery labor was billed as curative for its earthy, sensorial, and nurturing associations. "Prisoners from the penitentiary, familiar with sidewalks and tenements, learn how it feels to spade up moist earth and what happens to undesirable growth," a *New York Times* reporter wrote.[82] Prisoners tending trees were described as "flourishing";[83] physical bodies were improved by "getting back to nature"; and young offenders, the reporter highlighted, found new satisfaction in growing living things.[84] However, because of Rikers' short sentences and lack of sufficient support staff, some prison labor proponents worried that inmates left before they were properly "molded much by the influence of trees."[85]

Some inmates, speaking to journalists, have recalled the nursery detail, colloquially known as "going to the country," as being a welcome contrast from an otherwise overcrowded, miserable, and violent environment.[86] Few jobs on

Rikers involved time outside, and the facility walls blocked views to the island landscape and the surrounding skylines of the prisoners' home boroughs. "You're in a better place out here, you're not in jail, and you're not back at home, on the streets, getting into trouble either," seventeen-year-old Otis Williams, serving five months for shoplifting, told the *New York Times*. "You're in a kind of place you didn't know about."[87] Others expressed affinities for working with plants and a desire to work in horticulture afterwards. Corrections officers said that the nursery had fewer disciplinary problems because inmates wanted those specific jobs. "You get restless in jail," an inmate named Mitchell Parker told a reporter, "But you can get restless back in the world too, when you can't get a job and people are against you. At least here, [Parks Department Worker] Sam is on our side."[88]

While often promoted as a tool of reform, prison labor programs are motivated by profit. Even if the nursery was considered a preferable job to others, inmates worked uncompensated or under-compensated for their labor.[89] When the inmate population rebounded after World War II, one journalist joked about the tree nursery: "The situation is much better now [that there are more prison workers] – better for the trees, that is."[90] Inmate labor produced generation upon generation of healthy street trees, nurtured from seedlings to sturdy juveniles, but by not being paid prisoner labor ultimately made the operation sustainable, and it was lauded as a great success overall. While in many institutions it was customary to pay inmates a small token amount for their work, on Rikers in the 1950s, inmates weren't paid at all; by the 1980s nursery workers were said to make only 50 cents per hour – on the high end of the pay scale for Rikers workers.[91]

Although abolished alongside slavery in the 13th Amendment for the civilian population, forced and unpaid labor remains a legal punishment for convicted criminals. In *The New Jim Crow*, Michelle Alexander argues that incarceration and forced prison labor – not a program for reform – is the direct legacy of slavery.[92] With abolition, the loss of a large free labor force was catastrophic for the U.S. economy, and slavery was rapidly replaced with another violent mechanism – that of targeted incarceration and forced prison labor. Ava DuVernay's documentary film *13th* describes how this clause has been leveraged over a century and a half towards the disproportionate incarceration of urban residents of color.[93] A fair wage (at least minimum wage) was a central demand of the watershed 1971 Attica prison uprising in upstate New York, and it continues today as a central demand of nationwide prison labor strikes initiated on the forty-fifth anniversary of the uprising.[94] While the tree nursery on Rikers was a welcome experience for some inmates, it (and the trees) thrived because of its connection to this entrenched and unjust legacy.

By the late 1980s, the Rikers landscape had changed dramatically. The population had more than doubled in a decade,[95] and a converted Staten Island ferryboat was docked nearby to house the overflowing inmates.[96] Rikers grew as New York State's incarceration rates – which had been stable from 1880 to 1970 – skyrocketed following the implementation of the Rockefeller Drug Laws in 1973.[97] Introducing lengthy, mandatory sentences – equivalent to rape and assault terms – for non-violent possession and sale of drugs, the laws had much greater impact on New York City than elsewhere in the state. Military-style operatives focused their sweeps in predominantly black and Hispanic neighborhoods; Central Harlem was one of the communities disproportionately affected, and today it has the city's third-highest rate of incarceration. As epidemiologist Ernest Drucker describes it, the Rockefeller laws effectively "pumped" citizens into the unprecedented mass incarceration of America, expanding prisons and dismantling communities in their wake.[98] By the end of the 1980s, Rikers was over capacity and rapidly constructing new jails to house the swelling population.

"The nursery is about to become the city's latest crime victim," a journalist reported in a 1987 article subtitled, "Crushing Azaleas to Erect Jails."[99] "They're building new prisons and wiping out our nursery,"[100] twenty-year-old Pablo Jimenez, serving eight months, was quoted as saying. Despite the perceived success of the Rikers nursery, new jail facilities were planned to extend across the entire island, consuming the nursery land for their construction. Rikers Island's final two horticulturalists, Bob Zappala and Sam Bongiorno, worked overtime to move trees out of the way of the construction activity and get as many trees as possible shipped to the city streets. Zappala, a spirited storyteller now retired and doing freelance landscaping from Queens, worked at the Rikers nursery during its final eight years. When I met Bob at an Italian restaurant in Queens, he shared stories from his years working at the nursery. He had fond memories of working in the nursery, telling stories of some of the men that he worked with there and sharing photos. One black and white photo shows a few men working in the nursery, some digging by hand, another operating a small excavator. This and other photos mark the shrinking and closing of the tree nursery, all while the corrections operations expanded. Bob's job involved both closing the nursery down, and also setting up the replacement Citywide Nursery in Van Cortland Park in the Bronx.

For decades following the closure of the Rikers nursery, the complex has deteriorated. In 2014, a *New York Times* investigation revealed the wretched depths of abuse and harm happening on the island.[101] The same year, the Department of Justice announced its plan to sue New York City, calling out the facility for its culture of abuse and violence directed at youth.[102] Other reports have exposed disasters related to the land itself. The former landfill emits poisonous methane gas, surrounding waste transfer and power plant sites emit dangerous particulate, the whole island is vulnerable to flooding, and inmates suffer extreme heat and cold in the old, underserved facilities.[103] Community activist groups have been calling for the closure of Rikers for decades, and recently politicians and a wider group of proponents have joined these calls. A design competition sponsored by a business media company rendered plans showing the Rikers site reimagined as mixed-use, including a rehabilitation center, an extension of LaGuardia airport, and a new kind of park including, once again, an urban farm.[104] #CloseRikers, a movement initiated by JustLeadershipUSA and others, has been mobilizing support from elected officials and the public towards their aim. "Rikers is every jail and every jail is Rikers,"

Figure 4.19 Tree excavation, Rikers Island Nursery, 1987

Source: Photograph by Calvin Wilson (New York City Parks Photo Archive).

Figure 4.20 Robert Zappala, former Parks Department Rikers Island Nursery employee, 2015

said Glenn E. Martin of JustLeadershipUSA.[105] Rikers has become a symbol of the criminal justice system at large – corrupt, abusive, out of control – and the struggle to close it is part of the wider struggle for de-carceration, a new abolition. The complex – originally promoted as an institution of reform and rehabilitation – is so broken, they claim, that it must be shut down.

Unevenly Dappled Shade

Following the closure of the Rikers nursery, the city continued to operate municipal nurseries but shifted to private models, an arrangement that wasn't optimal for the Parks Department nor the commercial nurseries. The Parks Department had a hard time acquiring the species and numbers needed for large jobs, and they weren't able to get consistent quality from all of the nurseries. At the same time, commercial nurseries, in hopes of supplying these contracts, invested time and funds into growing large quantities of trees, but they had no assurance that the city would actually buy them. Today, the majority of New York City's future street and park trees grow in three private nurseries in Long Island, Chesapeake City (MD), and Buffalo.

On an early June visit to Whitman Nurseries in Long Island, Parks Department Director of Street Tree Planting Matthew Stephens and his team survey the rows of specified trees, counting specimens and making notes about how particular cultivars are doing. Matthew explains how the city now contracts growing to specific private nurseries on multi-year contracts, which finally match the multi-year timeline required to grow, tend, and prepare the trees exactly as needed, while guaranteeing purchases. They plan decades ahead, specifying numbers and species to grow, already having an idea of the streets and parks that will receive them, and ensure the preferred species and cultivars are growing. They can, for example, now request Kentucky coffee trees (*Gymnocladus dioicus*), robust city trees that nurseries just weren't growing before because they weren't common enough yet and considered a risky sell. Now they're planting 400 of them a year. Joe Sipala, Whitman's owner, says that this model makes the most sense from a horticultural standpoint. Working more closely with the Parks Department and their agendas, Sipala is able to try out new species and test lower-impact practices (by cutting back on chemical fertilizers and pesticides), knowing that the trees will meet the city's demands. It is a fairer and more efficient working relationship.

The acres upon acres of trees growing at Whitman Nurseries reflect the upsurge in street tree planting spawned by the Million Trees NY initiative, which put 1 million new trees in the five boroughs between 2007 and 2015. An influential report underscoring the health, environmental, and financial values of trees in the city was a primary motivation for the program. Like the Tree Planting Association had argued a century earlier, the report concluded that trees significantly temper undesirable environmental qualities, such as excess heat, air pollution, and energy consumption, as well as generally improve quality of life. Every year, across the city, street trees produce an estimated $121.9

Figure 4.21 **Rows of future New York City street trees, Whitman Nurseries, Riverhead, New York, 2015**

million in benefits. This figure includes $754,000 by sequestering and reducing carbon dioxide emissions; $5.27 million by removing, releasing, and avoiding pollutants; $35.6 million by intercepting rain and reducing storm water runoff; and $52.5 million by looking good and increasing property values. No longer a drain on resources, the report found that for every dollar spent, trees give about $5.60 back. The London plane, with its mature population and voluminous canopy, provides the most benefit, each tree providing $307 per year (compared to the average of $209.)[106] The longer the trees survive, the bigger they grow, and the more benefits they convey – all the more reason to commit to maintenance and ensure their survival.[107]

In Whitman's loading yard, dozens of balled and burlapped London plane trees await their two-hour truck journey to planting sites in New York City. Although still a much-loved local favorite, the London plane now accounts for just 4 percent of new plantings. Seventy years ago, the London plane was billed as the all-purpose tree hero, a hardy stock that could survive in the singularly tough urban environment. Today, rather than stick with a few species, the city now specifies more than 200; a wide variety of cultivars means less vulnerability to disease and greater ecological complexity. The "city" is no longer seen as one environment, a single concrete jungle in which every tree struggles to survive. Instead, a new planting protocol breaks urban environments into eighteen different "biotopes," distinguished by commercial or residential use, a gradient of toughness (indicating drought conditions, soil volume, and compaction), the width of the road, and the presence of overhead wires.[108] The protocol acknowledges the biophysical variation and nuances of urban environments, alongside the specific growth characteristics and habits of hundreds of different

Figure 4.22 **London plane trees, tagged for shipment,
Whitman Nurseries, Riverhead, New York, 2015**

tree species. No longer considered a fanciful or weak antidote to urbanization, street trees are now billed as protagonists of a green urban future, central characters in addressing the changing climate.[109]

In contrast to the vision of a future, unified "greener and greater city" for all, the current one is starkly unequal. While tree planting used to be thought of as a Parks special issue, Matthew Stephens tells me, trees are now commonly recognized as an equity issue. Historically uneven services, municipal investments, and systemic biases have led to an uneven pattern of tree-planting throughout the city boroughs. This unevenly dappled shade canopy illustrates a history of other inequities. Canopies flourish where more affluent residents enjoy better public services and open space amenities, and neighborhoods with fewer trees are correlated with poverty, worse air quality, and higher incidences of childhood asthma.[110] Trees are more likely to thrive in conditions where people also thrive – where sufficient services are in place, where environmental conditions support life. Where environmental conditions are harmful to people, trees could mediate their effects – but often, they aren't there in the first place, or suffer when they are. Researchers haven't fully parsed out causal relationships between trees, poverty, and health outcomes, but correlations are evident. Acknowledging these inequalities, the city now targets more than half of its new planting to priority areas, in an initiative called "Trees for Public Health." Of six identified areas with relatively few trees, dense populations, and significant health problems, East Harlem (north of 96th Street and west of Malcolm X Boulevard/Sixth Avenue) was deemed one of the most urgent.[111]

Continuous Care

As best as I can tell, the London plane tree that Hulan E. Jack ceremoniously planted in 1959 now towers over the sidewalk on Seventh Avenue and 117th Street in front of Esther Hair Salon. Sporting a lush canopy and graceful arching structure, it did, as Jack predicted back then, grow to be around 30 feet tall. On that block, three plane trees from that original planting remain, and one seems to have been replaced a decade or two ago. A few years after the 1959 planting, another 1,000 plane trees were planned to extend plantings on Seventh, Eighth, and Amsterdam Avenues from 110th up to 145th Street.[112] Community members and organizations participated in the upkeep and further planting of the street. Members of the Harlem Youth Opportunities Unlimited – Associated Community Team (HARYOU – ACT), a President Johnson–supported program to address rising unemployment and exacerbating social inequality, took on tree planting and landscaping in the area.[113] While the trees were planted and tended, residents continued to fight environmental and social crises that deepened over the coming decades. Private and municipal landlords began to abandon housing stock, and government began to disinvest further from

Figure 4.23 Mature London plane tree at Seventh Avenue and 117th Street, likely the same one planted in 1959, 2017

Source: © 2019 Google.

housing, social agencies, transportation, and infrastructure. As manufacturing jobs left the city, unemployment bottomed out in Central Harlem, and residents found themselves fighting a next level of disinvestment.

Today, the area around Seventh Avenue, with its grand apartment buildings, mature tree canopy, and proximity to Central Park, is an obvious pull for developers.[114] Up and down the street, in between the London plane trees, construction scaffolding signals upgrades and renovations. According to a 1984 study by Neil Smith and Richard Schaffer, Harlem was already experiencing the early stages of gentrification, which they called a "Catch-22" for residents.[115] Private investment brings much-needed improvements to the building stock and landscape, but this often leads to the displacement of the very residents who have worked and fought for decades to improve the vitality of the neighborhood. In the last twenty-five years, Central Harlem has seen the second-highest rate of rent increase in the city, yet its population has historically had one of the lowest average incomes, about half that of New York City's composite average, making the threat of displacement a pressing reality for many long-time residents.[116]

Ironically, beautiful new street trees, which in 1959 were a sign of optimism for a better future, can now be something to fear. Author and Harlem resident Michael Henry Adams tells the story of one local boy, taking in the messages of an anti-gentrification demonstration and telling his friends, "I told you they didn't plant those trees for us." Adams writes, "It was painful to realize how even a kid could see in every new building, every historic renovation, every boutique clothing shop – indeed in every tree and every flower in every park improvement – not a life-enhancing benefit but a harbinger of his own displacement."[117] Just as environmental justice puts the focus on the unequal distribution of environmental risks and benefits, the notion of "green gentrification" highlights the environmental injustices wrought in the name of urban sustainability.[118] A street tree might improve the quality of life for one resident, but its marketable green-lifestyle associations can also play a role in displacing another.

Over four decades, Rikers inmates planted, tended, pruned, watered, transplanted, dug out of the ground, and sent countless shipments of street trees, not to mention shrubs and ornamental trees, across the boroughs. While the Parks Department also purchased other trees from commercial nurseries, the official line was that Rikers supplied the entire city's trees.[119] Generations of Rikers trees landed in streets, sometimes with photo ops, as in the case of the 1959 Seventh Avenue planting. But more often, they were planted quietly and steadily, May through October, establishing their roots into pit after pit, street after street, year after year.

Artist Jill Hubley's online map of New York City street trees allows you to select certain species, zoom in and out, and navigate around the city.[120] You can zoom in and see the smattering of old and new London planes up and down Seventh Avenue. You can maneuver around and try to find the areas where large shipments of London plane trees from Rikers Island were planted in the 1940s through the 1980s: 1,600 along Sixth Avenue from Battery Park to Central Park;[121] 1,300 along seven miles of Third Avenue from Chatham Square to the Harlem River;[122] along West 87th Street between Central Park and Columbus Avenue;[123] on East 29th Street, between Third and Lexington;[124] and towards the newly constructed United Nations Secretariat along 42nd Street and Second Avenue.[125] You can zoom out and take in the 90,000 London planes living now throughout the boroughs. At this scale, if you turn on all the species, the page gets super saturated with color: a psychedelic montage of all species and ages lighting up the streets. It is impossible to know exactly how many of these trees, or which ones exactly, were grown in the nursery on Rikers Island, but a good number were. These trees experienced decades of attention and tending – by inmates, Parks Department workers, and community members – against certain odds. Cities bear many heroic individual development myths, but the voluminous tree canopy (and its invisible, undocumented, and uncompensated makers) reminds us of how a city really is a collective construction. Trees have stood as both an indicator of the health or illness in the urban environment, and now are more clearly understood as a partner of survival – in part because they need so much care.

Figure 4.24 **London plane trees in New York City, screenshot from Jill Hubley's digital mapping project, NYC Street Trees by Species, 2015**

Source: www.jillhubley.com/project/nyctrees/#PLAC.

Notes

1 Florence McDowell, "Whatsoever Things are Lovely," *The Crisis: A Record of the Darker Races* 51, no. 5 (1944): 160.

2 Augustine Henry and Margaret G. Flood, "The History of the London Plane, *Platanus Acerifolia*, with notes on the Genus Platanus," *Proceedings of the Royal Irish Academy. Section B: Biological, Geological, and Chemical* Science 35 (1919/1920): 9.

3 Michelle H. Martin and P.J. Coughtrey, *Biological Monitoring of Heavy Metal Pollution: Land and Air* (London: Applied Science Publishers, 1982), 116.

4 Kajetan Dzierżanowski, Robert Popek, Helena Gawrońska, Arne Sæbø, and Stanislaw W. Gawroński, "Deposition of Particulate Matter of Different Size Fractions on Leaf Surfaces and in Waxes of Urban Forest Species," *International Journal of Phytoremediation* 13, no. 10 (2011).

5 Thomas J. Campanella, "Landscape Architecture: The Roman Roots of Gotham's London Plane," *Wall Street Journal*, July 20, 2011, D5.

6 New York City Department of Parks, "Press Release," August 26, 1948.

7 Max Page, *The Creative Destruction of Manhattan, 1900–1940* (Chicago: The University of Chicago Press, 2001), 179.

8 H.R. Francis, "Report on the Street Trees of the City of New York," for the *Tree Planting Association of New York City*, Bulletin 15, no. 1e (Syracuse: The New York State College of Forestry, 1914), 17.

9 Jacob Joshua Levison, *Improve Your Street: Plant Trees Get Your Neighbors to Cooperate* (Brooklyn: American Association for the Planting and Preservation of City Trees, 1912), 2.

10 "Tree Planting in New York," *New York Times*, April 29, 1902, 8.

11 Ibid.

12 Page, *The Creative Destruction of Manhattan*, 202.

13 For a thorough account of the Tree Planting Association's work, see Sonja Dümpelmann, *Seeing Trees: A History of Street Trees in New York City and Berlin* (New Haven: Yale University Press, 2019).

14 Stephen Smith, "Vegetation a Remedy for the Summer Heat of Cities," *Appleton's Popular Science Monthly* 54 (1899): 433.

15 "New York is Being Surveyed for Planting of Trees," *New York Times Magazine*, January 4, 1914.

16 Ibid.

17 William Solotaroff, *Shade-Trees in Towns and Cities: Their Selection, Planting, and Care as Applied to the Art of Street Decoration; Their Diseases and Remedies, Their Control and Supervision* (New York: John Wiley & Sons, 1911), 82–8.

18 "Want City Bureau for Tree Culture," *New York Times*, March 15, 1914, 4.

19 Laurie Davidson Cox, *A Street Tree System for New York City, Borough of Manhattan* (Syracuse: New York State College of Forestry, 1916), 24.

20 Jonathan Gill, *Harlem: The Four Hundred Year History From Dutch Village to Capital of Black America* (New York: Open Road + Grove/Atlantic, 2011), 105.

21 David Maurrasse, *Listening to Harlem: Gentrification, Community, and Business* (New York: Routledge, 2006), 17.

22 Ibid., 18.

23 Jay Shockley, "Landmarks Preservation Commission, Designation List 172, LP-1254," *New York*, October 16, 1984, 8.

24 Gilbert Osofsky, *Harlem: The Making of a Ghetto, Negro New York 1890–1930* (New York: Harper & Row Publishers, 1963), 130.

25 Maurrasse, *Listening to Harlem*, 18–20.

26 Ibid., 14.

27 Isabel Wilkerson, *The Warmth of Other Suns: The Epic Story of America's Great Migration* (New York: Vintage, 2011), 161.

28 Gill, *Harlem*, 173.

29 Maurasse, *Listening to Harlem*, 20.

30 Wallace Thurman, *Negro Life in New York's Harlem: A Lively Picture of a Popular and Interesting Section* (Girard: Haldeman-Julius Publications, 1927), 8–10.

31 Kareem Abdul-Jabbar, *On the Shoulders of Giants: My Journey Through the Harlem Renaissance* (New York: Simon and Schuster, 2007), 28.

32 Sam Roberts, *A History of New York in 101 Objects* (New York: Simon and Schuster, 2014), 171.

33 Maurrasse, *Listening to Harlem*, 22.

34 Stephen Robertson, "Parades in 1920's Harlem," *Digital Harlem Blog*, February 1, 2011, https://digitalharlemblog.wordpress.com/2011/02/01/parades-in-1920s-harlem.

35 Ibid.

36 Shannon King, *Whose Harlem Is This, Anyway? Community Politics and Grassroots Activism during the New Negro Era* (New York: NYU Press, 2015), 188–9.

37 Maurasse, *Listening to Harlem*, 18.

38 Ibid., 24.

39 Osofsky, *Harlem*, 141–3.

40 Charles G. Bennett, "'Shocking' Housing Hazards in Harlem Found in City Tour," *New York Times*, October 21, 1952, 1.

41 "Harlem Beautiful Plan is Announced," *New York Amsterdam News*, April 13, 1946, 7.

42 "Should 7th Ave. Be Carver Blvd.?" *New York Amsterdam News*, July 21, 1951, 1.

43 "Harlem's Beautification Program Runs into a Snag," *New York Amsterdam News*, July 27, 1946, 2.

44 "7th Avenue to Get Real Facelifting," *New York Amsterdam News*, February 7, 1959, 1.

45 Elmer A. Carter, "Suggestion for Improvement," *New York Amsterdam News*, October 2, 1943, 12.

46 "War Against Rats Pushed in Harlem," *New York Times*, June 29, 1952, 33.

47 Jane Boutwell and Brendan Gill, "Parisian," The Talk of the Town, *The New Yorker*, January 19, 1957, 21.

48 "Jack to Give 7th Avenue New Lights," *New York Amsterdam News*, February 28, 1959, 1.

49 "7th Avenue to Get Real Facelifting," 1.

50 Ibid.

51 "Uptown 7th Ave. Dons Easter Trees," *New York Times*, March 17, 1959, 29.

52 John Walker Harrington, "$10,000,000 City Penitentiary, Rising on Riker's Island in East River," *New York Herald Tribune*, July 19, 1931, A6.

53 Ibid.

54 "Rikers Island Project to End Dumping Nov. 1," *New York Herald Tribune*, September 2, 1942, 20.

55 Harrington, "$10,000,000 City Penitentiary, Rising on Riker's Island in East River," A6.

56 "Convicts to Raise City Shade Trees," *New York Times*, August 23, 1939, 1.

57 New York City Department of Parks, "1941 Press Releases Part 2," Arsenal, Central Park, December 29, 1941, http://home2.nyc.gov/html/records/pdf/govpub/41971941_press_releases_part2.pdf.

58 "Convicts to Raise City Shade Trees," *New York Times*, August 23, 1939, 1.

59 "Rikers Island Dumps Yield to Tree Nursery," *New York Herald Tribune*, October 6, 1940, A1.

60 R.A. Ogan, "Memo: Street Tree Planting – Capital Budget Program," March 12, 1946, City of New York Department of Records and Information Services Municipal Archives, Box 102719 Folder 53.

61 New York City Department of Parks, "1941 Press Releases Part 2".

62 "Rikers Island Nursery to Add 10,000 Trees," *New York Herald Tribune*, May 24, 1942, C6.

63 "Convicts to Raise City Shade Trees," *New York Times*, August 23, 1939, 1.

64 Ken Clarke, "Trees in the Concrete Forest," *New York Herald Tribune*, April 22, 1951, SM20.

65 "Island Nursery," *New Yorker*, June 28, 1947, 18.

66 Ken Clarke, "Trees in the Concrete Forest," SM20.

67 "Rikers Island Dumps Yield to Tree Nursery," A1; "Riker's Island Nursery," New York City Department of Parks, Press Release, City of New York Department of Records and Information Services Municipal Archives, Box 102651, Folder 009, 1943.

68 New York City Department of Parks, "Correspondence From Cornelius M. O'Shea to Mr. H.L. Li," April 22, 1956, City of New York Department of Records and Information Services Municipal Archives, Box 102923, Folder 018.

69 New York City Department of Parks, "Cost &, Propagation of Trees at Rikers Island," Correspondence From C. M. O'Shea, to Stuart Constable, October 22, 1956; "Riker's Island Nursery," City of New York Department of Records and Information Services Municipal Archives, Box 102923, Folder 018, 1956.

70 New York City Department of Parks, "Rikers Island Nursery."

71 New York City Department of Parks, "Replacement of Dead Trees Along Lower East River Drive," Correspondence From Arthur S. Hodgkiss to Commissioner Moses, March 31, 1944, and "Riker's Island Equipment, Nursery," 1944, City of New York Department of Records and Information Services Municipal Archives, Box 102679, Folder 045, Department of Parks & Recreation, 1944.

72 Philip Benjamin, "Tree-Moving Set on Riker's Island: 15,000 in Nursery to Be Replaced by Buildings," *New York Times*, September 17, 1964, 39.

73 "Welcome to Young Trees," *New York Herald Tribune*, November 4, 1950, 10.

74 William E. Geist. "On Rikers Island: Crushing Azaleas to Erect Jails," *New York Times*, April 25, 1987, 33.

75 "Island Nursery," 18.

76 "Trees," *The New Yorker*, October 7, 1944, 14.

77 New York City Department of Parks, "Riker's Island Nursery," Correspondence From C.M. O'Shea to Sam M. White, February 15, 1956, City of New York Department of Records and Information Services Municipal Archives, Box 102923, Folder 018.

78 New York City Department of Parks, "Trees & Shrubs Destroyed at Rikers Island Park Department Nursery," Correspondence From C.M. O'Shea to John Collins, February 5, 1957, City of New York Department of Records and Information Services Municipal Archives, Folder 24, Box 102950.

79 Richard J. H. Johnston, "2 Jailbreakers Seized after 2 Days in a Foxhole," *New York Times*, June 16, 1966, 38.

80 Benjamin, "Tree-moving Set on Rikers Island," 39.

81 "Rikers Island Nursery to Add 10,000 Trees," *New York Herald Tribune*, C6.

82 Bernard Stengren, "Third Avenue Trees," *New York Times*, April 1, 1956, 204.

83 Ibid.

84 William M. Farrell, "'Rustic 'Paradise' on Rikers Island," *New York Times*, April 29, 1953, 31.

85 "Trees," *The New Yorker*, October 7, 1944, 14.

86 Geist, "On Rikers Island, 33.

87 Ibid.

88 Ibid.

89 Raymond Price Jr., "Dilemma at City Prison," *New York Herald Tribune*, April 25, 1959, 10.

90 "Park Department Puts Out 2,100 New Trees," *New York Times*, May 7, 1947, 29.

91 Clyde Haberman, "City Is Using Rikers Prisoners to Clean Parks," *New York Times*, January 21, 1982, 1.

92 Michelle Alexander, *The New Jim Crow: Mass Incarceration in the Age of Color-blindness* (New York: The New Press, 2012).

93 Ava Duvernay and Jason Moran. *13TH*. USA, 2016.

94 Juleyka Lantigua-Williams, "Is Another Attica Uprising on the Horizon?" *The Atlantic*, September 9, 2016, www.theatlantic.com/politics/archive/2016/09/is-another-attica-on-the-horizon/499397.

95 "Pay Prisoners for Overcrowding, Federal Judge Tells New York," *Chicago Tribune*, November 30, 1990, D24.

96 Geist, "On Rikers Island," 33.

97 Ernest Drucker, *A Plague of Prisons: The Epidemiology of Mass Incarceration in America* (New York: The New Press, 2011), 51. Note that some community organizations also supported these laws, including Glester Hinds of the Peoples Civic and Welfare Association, see Michael Javen Fortner, *Black Silent Majority: The Rockefeller Drug Laws and the Politics of Punishment* (Cambridge: Harvard University Press, 2015), 1.

98 Drucker, *A Plague of Prisons*, 55.

99 Geist, "On Rikers Island," 33.

100 Ibid.

101 Michael Winerip and Michael Schwirtz, "Where Mental Illness Meets Brutality in Jail," *The New York Times*, July 14, 2014, www.nytimes.com/2014/07/14/nyregion/rikers-study-finds-prisoners-injured-by-employees.html?mtrref=undefined&gwh=13 8C2162F5C8E40CB25E7C538F42E1AC&gwt=pay&assetType=nyt_now.

102 Southern District of New York, U.S Attorneys Office, "Department of Justice Takes Legal Action to Address Pattern And Practice of Excessive Force and Violence At Rikers Island Jails that Violates the Constitutional Rights Of Young Male Inmates," December 18, 2014, www.justice.gov/usao-sdny/pr/department-justice-takes-legal-action-address-pattern-and-practice-excessive-force-and (see also www.npr.org/2014/12/18/371721082/justice-department-sues-over-conditions-at-rikers-island-jail); Melissa Block, "Justice Department Sues Over Conditions in Rikers Island Jail," Audio Podcast, *NPR: All Things Considered*, December 18, 2014, www.npr.org/2014/12/18/371721082/justice-department-sues-over-conditions-at-rikers-island-jail.

103 Raven Rakia, "A Sinking Jail: The Environmental Disaster That Is Rikers Island," *Grist*, March 15, 2016, http://grist.org/justice/a-sinking-jail-the-environmental-disaster-that-is-rikers-island.

104 See proposal by Architecture firm Perkins + Will, in: Aaron Elstein and Joe Anuta, "Rikers Reimagined: Innovative Ideas to Turn the Infamous Island into a New York Destination," *Crains New York*, February 27, 2016, www.crainsnewyork.com/article/20160228/REAL_ESTATE/160229896/innovative-ideas-to-turn-the-infamous-rikers-island-into-a-new-york-destination.

105 Lantigua-Williams, "Can a Notorious New York City Jail Be Closed?"

106 Paula J. Peper, E. Gregory McPherson, James R. Simpson, Shelley L. Gardner, Kelaine E. Vargas, and Qingfu Xiao, "New York City, New York Municipal Forest Resource Analysis," *USDA Forest Service, Pacific Southwest Research Station* (Center for Urban Forest Research), 2007, 2, 3, 21, www.milliontreesnyc.org/downloads/pdf/nyc_mfra.pdf.

107 Jacqueline Lu, Morgan Monaco, Andrew Newman, Ruth A. Rae, Lindsay K. Campbell, Nancy Flaxa-Raymond, and Erika S. Svendsen, *Million Trees NYC: The Integration of Research and Practice* (NYC Parks and USDA Forest Services), 23, www.milliontreesnyc.org/downloads/pdf/MTNYC_Research_Spreads.pdf.

108 David Moore, "A New Method for Streamlining Tree Selection in New York City," *New York City Parks Department*, www.nycgovparks.org/pagefiles/82/streamlining-tree-selection-in-nyc.pdf, 2015.

109 Lu et al., *Million Trees NYC: The Integration of Research and Practice*.

110 Gina S. Lovasi, Jarlath P.M. O'Neil-Dunne, Jacqueline W.T. Lu, Daniel Sheehan, Matthew S. Perzanowski, Sean W. MacFaden, Kristen L. King, Thomas Matte, Rachel L. Miller, Lori A. Hoepner, Frederica P. Perera, and Andrew Rundle "Urban Tree Canopy and Asthma, Wheeze, Rhinitis, and Allergic Sensitization to Tree Pollen in a New York City Birth Cohort," *Environmental Health Perspectives*, January 15, 2013, http://dx.doi.org/10.1289/ehp.1205513.

111 Lu et al., *Million Trees NYC, Neighborhoods*, www.milliontreesnyc.org/html/million_trees/neighborhoods.shtml.

112 "Trees Sprouting up in Harlem," *New York Amsterdam News*, April 6, 1963, 6.

113 "2500 Jobs for Kids in Harlem," *New York Amsterdam News*, June 12, 1965, 1.

114 Maurasse, *Listening to Harlem*, 6.

115 Richard Schaffer and Neil Smith, "The Gentrification of Harlem?" *Annals of the Association of American Geographers* 76, no. 3 (1986): 362.

116 Furman Center New York University, *Focus on Gentrification*, New York: NYU Furman Center, 2016, http://furmancenter.org/files/sotc/Part_1_Gentrification_SOCin2015_9JUNE2016.pdf.

117 Michael Henry Adams, "The End of Black Harlem," *The New York Times*, May 29, 2016, www.nytimes.com/2016/05/29/opinion/sunday/the-end-of-black-harlem.html.

118 Kenneth A. Gould and Tammy L. Lewis, *Green Gentrification: Urban Sustainability and the Struggle for Environmental Justice* (New York: Routledge, 2017), 153.

119 "Tree-Planting Project of City in Full Swing," *New York Herald Tribune*, November 1, 1950, 25.

120 Jill Hubley, "New York City Street Trees by Species," *Jill Hubley* (blog), April 11, 2015, http://jillhubley.com/blog/nyctrees; New York City Open Data, "2015 Street Tree Census – Tree Data," https://data.cityofnewyork.us/Environment/2015-Street-Tree-Census-Tree-Data/pi5s-9p35.

121 New York City Department of Parks, Correspondence from Robert Moses to Mr. Edward H. Scanlon, February 16, 1945, City of New York Department of Records and Information Services Municipal Archives. Box 102691, Folder 42.

122 Stengren, "Third Avenue Trees," 204; "Prettying Up Third Ave.," *New York Herald Tribune*, April 6, 1956, A1.

123 "West Side Tree Project Extended," *New York Times*, December 4, 1955, 54.

124 New York City Department of Parks, Correspondence from John A. Mulcahy to Mr. George W. Thompson, November 22, 1957, City of New York Department of Records and Information Services Municipal Archives, Box 102950, Folder 25.

125 New York City Department of Parks, City of New York Department of Records and Information Services Municipal Archives, Box 102981, Folder 20, 1957.

Bibliography

"2500 Jobs for Kids in Harlem." *New York Amsterdam News*, June 12, 1965.

"7th Avenue to Get Real Facelifting." *New York Amsterdam News*, February 7, 1959.

Abdul-Jabbar, Kareem. *On the Shoulders of Giants: My Journey Through the Harlem Renaissance*. New York: Simon and Schuster, 2007.

Adams, Michael Henry. "The End of Black Harlem." *The New York Times*, May 29, 2016. www.nytimes.com/2016/05/29/opinion/sunday/the-end-of-black-harlem.html.

Alexander, Michelle. *The New Jim Crow: Mass Incarceration in the Age of Colorblindness*. New York: The New Press, 2012.

Benjamin, Philip. "Tree-Moving Set on Rikers Island: 15,000 in Nursery to be Replaced by Buildings." *New York Times*, September 17, 1964.

Bennett, Charles G. "'Shocking' Housing Hazards in Harlem Found in City tour." *New York Times*, October 21, 1952.

Block, Melissa. "Justice Department Sues Over Conditions in Rikers Island Jail." Audio Podcast, *NPR: All Things Considered*, December 18, 2014, www.npr.org/2014/12/18/371721082/justice-department-sues-over-conditions-at-rikers-island-jail.

Boutwell, Jane and Brendan Gill. "Parisian – The Talk of the Town." *The New Yorker*, January 19, 1957.

Campanella, Thomas J. "Landscape Architecture: The Roman Roots of Gotham's London Plane." *Wall Street Journal*, July 20, 2011, D5.

Carter, Elmer A. "Suggestion for Improvement." *New York Amsterdam News*, October 2, 1943.

Clarke, Ken. "Trees in the Concrete Forest." *New York Herald Tribune*, April 22, 1951.

"Convicts to Raise City Shade Trees." *New York Times*, August 23, 1939.

Cox, Laurie Davidson. *A Street Tree System for New York City, Borough of Manhattan*. Syracuse: New York State College of Forestry, 1916.

Drucker, Ernest. *A Plague of Prisons: The Epidemiology of Mass Incarceration in America*. New York: The New Press, 2011.

Dümpelmann, Sonja. *Seeing Trees: A History of Street Trees in New York City and Berlin*. New Haven: Yale University Press, 2019.

Duvernay, Ava, and Jason Moran. *13TH*. USA, 2016.

Dzierżanowski, Kajetan, Robert Popek, Helena Gawrońska, Arne Sæbø, and Stanislaw W. Gawroński. "Deposition of Particulate Matter of Different Size Fractions on Leaf Surfaces and in Waxes of Urban Forest Species." *International Journal of Phytoremediation* 13, no. 10 (2011): 1037–46.

Elstein, Aaron, Joe Anuta, and Rosa Goldensohn. "Rikers Reimagined: Innovative Ideas to Turn the Infamous Island into a New York Destination." *Crains New York*, February 27, 2016, www.crainsnewyork.com/article/20160228/REAL_ESTATE/160229896/innovative-ideas-to-turn-the-infamous-rikers-island-into-a-new-york-destination.

Farrell, William M. "Rustic 'Paradise' on Rikers Island." *New York Times*, April 29, 1953.

Fortner, Michael Javen. *Black Silent Majority: The Rockefeller Drug Laws and the Politics of Punishment*. Cambridge, MA: Harvard University Press, 2015.

Francis, H.R. "Report on the Street Trees of the City of New York." *For the Tree Planting Association of New York City*. Bulletin 15, no. 1e. Syracuse: The New York State College of Forestry, 1914.

Furman Center New York University. *Focus on Gentrification*. New York: NYU Furman Center, 2016, http://furmancenter.org/files/sotc/Part_1_Gentrification_SOCin2015_9JUNE2016.pdf.

Geist, William E. "On Rikers Island: Crushing Azaleas to Erect Jails." *New York Times*, April 25, 1987.

Gill, Jonathan. *Harlem: The Four Hundred Year History From Dutch Village to Capital of Black America*. New York: Open Road + Grove/Atlantic, 2011.

Gould, Kenneth A., and Tammy L. Lewis. *Green Gentrification: Urban Sustainability and the Struggle for Environmental Justice*. New York: Routledge, 2017.

Haberman, Clyde. "City Is Using Rikers Prisoners to Clean Parks." *New York Times*, January 21, 1982.

"Harlem Beautiful Plan Is Announced." *New York Amsterdam News*, April 13, 1946.

"Harlem's Beautification Program Runs into a Snag." *New York Amsterdam News*, July 27, 1946.

Harrington, John Walker. "$10,000,000 City Penitentiary, Rising on Riker's Island in East River." *New York Herald Tribune*, July 19, 1931.

Henry, Augustine and Margaret G. Flood. "The History of the London Plane, Platanus Acerifolia, with notes on the Genus Platanus." *Proceedings of the Royal Irish Academy. Section B: Biological, Geological, and Chemical Science* 35 (1919/1920): 9–28.

Hubley, Jill. "New York City Street Trees by Species." *Blog*, April 11, 2015, http://jill-hubley.com/blog/nyctrees

Island Nursery." *New Yorker*, June 28, 1947.

Jack to Give 7th Avenue New Lights." *New York Amsterdam News*, February 28, 1959.

Johnston, Richard J.H. "2 Jailbreakers Seized after 2 Days in a Foxhole." *New York Times*, June 16, 1966.

King, Shannon. *Whose Harlem Is this, Anyway? Community Politics and Grassroots Activism during the New Negro Era*. New York: NYU Press, 2015.

Lantigua-Williams, Juleyka. "Can a Notorious New York City Jail be Closed?" *The Atlantic*, April 26, 2016. www.theatlantic.com/politics/archive/2016/04/will-rikers-island-be-closed/479790/.

Levison, Jacob Joshua. *Improve Your Street: Plant Trees Get Your Neighbors to Cooperate*. Brooklyn: American Association for the Planting and Preservation of City Trees, 1912.

Lovasi, Gina S., Jarlath P.M. O'Neil-Dunne, Jacqueline W.T. Lu, Daniel Sheehan, Matthew S. Perzanowski, Sean W. MacFaden, Kristen L. King, Thomas Matte, Rachel L. Miller, Lori A. Hoepner, Frederica P. Perera, and Andrew Rundle. "Urban Tree Canopy and Asthma, Wheeze, Rhinitis, and Allergic Sensitization to Tree Pollen in a New York City Birth Cohort." *Environmental Health Perspectives*, January 15, 2013, http://dx.doi.org/10.1289/ehp.1205513.

Lu, Jacqueline, Morgan Monaco, Andrew Newman, Ruth A. Rae, Lindsay K. Campbell, Nancy Flaxa-Raymond, and Erika S. Svendsen. *Million Trees NYC: The Integration of Research and Practice*. NYC Parks and USDA Forest Services, www.milliontreesnyc.org/downloads/pdf/MTNYC_Research_Spreads.pdf.

Martin, Michelle H., and P.J. Coughtrey. *Biological Monitoring of Heavy Metal Pollution: Land and Air*. Netherlands: Springer Science + Business Media, 2012.

Maurrasse, David. *Listening to Harlem: Gentrification, Community, and Business*. New York: Routledge, 2006.

McDowell, Florence. "Whatsoever Things Are Lovely." *The Crisis: A Record of the Darker Races* 51, no. 5 (1944): 160–72.

Moore, David. "A New Method for Streamlining Tree Selection in New York City." *New York City Parks Department*, 2015, www.nycgovparks.org/pagefiles/82/streamlining-tree-selection-in-nyc.pdf.

New York City Department of Parks. "City of New York Department of Records and Information Services Municipal Archives." 1957, Box 102981, Folder 20.

New York City Department of Parks. "Correspondence From Cornelius M. O'Shea to Mr. H.L. Li." April 22, 1956, City of New York Department of Records and Information Services Municipal Archives, Box 102923, Folder 018.

New York City Department of Parks. "Correspondence From John A. Mulcahy to Mr. George W. Thompson." November 22, 1957, City of New York Department of Records and Information Services Municipal Archives, Box 102950, Folder 25.

New York City Department of Parks. "Correspondence From Robert Moses to Mr. Edward H. Scanlon." February 16, 1945, City of New York Department of Records and Information Services Municipal Archives. Box 102691, Folder 42.

New York City Department of Parks. "Cost &, Propagation of Trees at Rikers Island." Correspondence From C. M. O'Shea, to Stuart Constable, October 22, 1956.

New York City Department of Parks. "Press Release." August 26, 1948, City of New York Department of Records and Information Services Municipal Archives.

New York City Department of Parks. "Press Release." City of New York Department of Records and Information Services Municipal Archives, Box 102651, Folder 009, 1943.

New York City Department of Parks. "Press Releases Part 2." *Arsenal*, Central Park, December 29, 1941, http://home2.nyc.gov/html/records/pdf/govpub/41971941_press_releases_part2.pdf.

New York City Department of Parks. "Replacement of Dead Trees Along Lower East River Drive." Correspondence From Arthur S. Hodgkiss to Commissioner Moses, March 31, 1944. City of New York Department of Records and Information Services Municipal Archives.

New York City Department of Parks. "Riker's Island Equipment, Nursery." City of New York Department of Records and Information Services Municipal Archives, Box 102679, Folder 045, 1944.

New York City Department of Parks. "Riker's Island Nursery." City of New York Department of Records and Information Services Municipal Archives, Box 102923, Folder 018, 1956.

New York City Department of Parks. "Riker's Island Nursery." Correspondence From C.M. O'Shea to Sam M. White, February 15, 1956, City of New York Department of Records and Information Services Municipal Archives, Box 102923, Folder 018.

New York City Department of Parks. "Trees & Shrubs Destroyed at Rikers Island Park Department Nursery." Correspondence From C.M. O'Shea to John Collins, February 5, 1957, City of New York Department of Records and Information Services Municipal Archives, Box 102950, Folder 24.

New York City Open Data. "2015 Street Tree Census – Tree Data." https://data.cityof newyork.us/Environment/2015-Street-Tree-Census-Tree-Data/pi5s-9p35.

"New York Is Being Surveyed for Planting of Trees." *New York Times*, January 4, 1914.

Ogan, R.A. "Memo: Street Tree Planting – Capital Budget Program." *City of New York Department of Records and Information Services Municipal Archives*, March 12, 1946, Box 102719, Folder 53 (New York City Department of Parks).

Osofsky, Gilbert. *Harlem: The Making of a Ghetto, Negro New York 1890–1930*. New York: Harper & Row Publishers, 1963.

Page, Max. *The Creative Destruction of Manhattan, 1900–1940*. Chicago: The University of Chicago Press, 2001.

"Park Department Puts Out 2,100 New Trees." *New York Times*, May 7, 1947.

"Pay Prisoners for Overcrowding, Federal Judge Tells New York." *Chicago Tribune*, November 30, 1990.

Peper, Paula J., E. Gregory McPherson, James R. Simpson, Shelley L. Gardner, Kelaine E. Vargas, and Qingfu Xiao, "New York City, New York Municipal Forest Resource Analysis." *USDA Forest Service, Pacific Southwest Research Station* (Center for Urban Forest Research), 2007, 2. Accessed December 21, 2016, www.milliontreesnyc. org/downloads/pdf/nyc_mfra.pdf.

"Prettying Up Third Ave." *New York Herald Tribune*, April 6, 1956.

Price Jr., "Raymond. "Dilemma at City Prison." *New York Herald Tribune*, April 25, 1959.

Rakia, Raven. "A Sinking Jail: The Environmental Disaster that Is Rikers Island." *Grist*, March 15, 2016, http://grist.org/justice/a-sinking-jail-the-environmental-disaster-that-is-rikers-island.

"Rikers Island Dumps Yield to Tree Nursery." *New York Herald Tribune*, October 6, 1940.

"Rikers Island Nursery to Add 10,000 Trees." *New York Herald Tribune*, May 24, 1942.

"Rikers Island Project to End Dumping Nov. 1." *New York Herald Tribune*, September 2, 1942.

Roberts, Sam. *A History of New York in 101 Objects*. New York: Simon and Schuster, 2014.

Robertson, Stephen. "Parades in 1920's Harlem." *Digital Harlem Blog*, February 1, 2011, https://digitalharlemblog.wordpress.com/2011/02/01/parades-in-1920s-harlem/.

Schaffer, Richard and Neil Smith. "The Gentrification of Harlem?" *Annals of the Association of American Geographers* 76, no. 3 (1986).

Shockley, Jay. "Landmarks Preservation Commission, Designation List 172, LP-1254." *New York*, October 16, 1984.

"Should 7th Ave. be Carver Blvd.?" *New York Amsterdam News*, July 21, 1951.

Smith, Stephen. "Vegetation a Remedy for the Summer Heat of Cities." *Appleton's Popular Science Monthly* 54 (1899): 433–50.

Solotaroff, William. *Shade-Trees in Towns and Cities: Their Selection, Planting, and Care as Applied to the Art of Street Decoration; Their Diseases and Remedies, Their Control and Supervision.* New York: John Wiley & Sons, 1911.

Southern District of New York, U.S Attorneys Office. "Department of Justice Takes Legal Action to Address Pattern and Practice of Excessive Force and Violence at Rikers Island Jails that Violates the Constitutional Rights of Young Male Inmates." December 18, 2014, www.justice.gov/usao-sdny/pr/department-justice-takes-legal-action-address-pattern-and-practice-excessive-force-and (see also www.npr.org/2014/12/18/371721082/justice-department-sues-over-conditions-at-rikers-island-jail).

Stengren, Bernard. "Third Avenue Trees." *New York Times*, April 1, 1956.

Thurman, Wallace. *Negro Life in New York's Harlem: A Lively Picture of a Popular and Interesting Section.* Girard: Haldeman-Julius Publications, 1927.

"Tree Planting in New York." *New York Times*, April 29, 1902.

"Tree-Planting Project of City in Full Swing." *New York Herald Tribune*, November 1, 1950.

"Trees Sprouting Up in Harlem." *New York Amsterdam News*, April 6, 1963.

"Trees." *The New Yorker*, October 7, 1944.

"Uptown 7th Ave. Dons Easter Trees." *New York Times*, March 17, 1959.

"Want City Bureau for Tree Culture." *New York Times*, March 15, 1914.

"War Against Rats Pushed in Harlem." *New York Times*, June 29, 1952.

"Welcome to Young Trees." *New York Herald Tribune*, November 4, 1950.

"West Side Tree Project Extended." *New York Times*, December 4, 1955.

Wilkerson, Isabel. *The Warmth of Other Suns: The Epic Story of America's Great Migration.* Reprint edition. New York: Vintage, 2011.

Winerip, Michael and Michael Schwirtz. "Where Mental Illness Meets Brutality in Jail." *The New York Times*, July 14, 2014, www.nytimes.com/2014/07/14/nyregion/rikers-study-finds-prisoners-injured-by-employees.html?mtrref=undefined&gwh=138C2162F5C8E40CB25E7C538F42E1AC&gwt=pay&assetType=nyt_now.

Figure 5.1(A) Detail, ipe logs and boards, sawmill, Belém, Brazil, 2012

Figure 5.1(B) Detail, Tenth Avenue Square, the High Line

Source: Photograph by Cynthia Goodson, 2013.

Chapter 5

Arresting Decay

Tropical Hardwood from Para, Brazil, to the High Line, 2009

On a tree where some bark has been torn off, the outer sapwood is exposed to the air. Right there, just underneath the bark, so much is going on: the cambium layer divides to make an outer ring of phloem cells and an inner ring of xylem cells. The phloem cells draw sugars from photosynthesizing leaves down into woody tissues and roots, and the xylem cells pull water and minerals up through the tree like straws. Here, new layers of xylem are added annually, marking the changing seasons with darkened growth rings, and thickening and lengthening the tree. The sapwood, you could say, is where all the living happens. Each year the inner layers of sapwood transform into the tree's core heartwood, adding more structure to the tree but also – physiologically speaking – dying.

Airborne fungal spores fall and land on the same exposed sapwood. Encouraged by just the right amount of moisture, temperature, and easy access to the cells the spores germinate into hyphae, the flossy threads that multiply into mycelium masses. As they bore holes deeper into the tree's core, their enzymes decompose the complex molecules that structure the cell walls, cellulose, and lignin, far ahead of the hyphae's location. In some cases, this decay might spout fruiting bodies – mushrooms easily spotted by the human eye. Brown-rot fungi (drawn more to softwoods) consume cellulose alone, leaving behind dark, crumbly, cubic material that can be crushed between your fingers. Whereas white-rot fungi (drawn more to hardwoods) consume both cellulose and lignin, leaving behind a bleached apparition of spongy wood.[1] The word "decay" is both a noun describing the physical condition of rotted wood, as well as a verb describing the "external digestion" of a tree, including the complex of microorganisms, chemical compounds, and stages of physical change to the wood.[2] Also known as "rot," "dote," "doze," or "punk,"[3] decay is behind the most sumptuous sylvan senses: the pungent smells and spongy textures that signal perpetual growth and decomposition.

Living sapwood cells actively resist fungi. In a process called compartmentalization, the cells surround incoming fungi and create a protective layer to stop it from spreading.[4] In living trees, the sapwood tends to be more resistant to fungi, whereas the internal heartwood is more vulnerable. But in wood products, the reverse is true: sapwood has less resistance, while the denser heartwood tends to be more resistant to decay. While some species are known

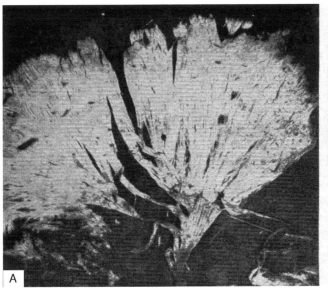

Figure 5.2 (A) Mycelium on stored wood and (B) mycelium of *Trametes pini* in southern yellow pine; the small black lines are fungus threads

Source: Otto Kress, C. J. Humphrey, C.A. Richards, M.W. Bray, J.A. Staidl, *Control of Decay in Pulp and Pulp Wood. Bulletin 1298.* United States Department of Agriculture, April 1925, Plate I.

to have greater natural decay resistance than others, individuals' resistance depends on their growing and moisture conditions and the types of nearby fungus.[5] As a whole, the forest ecosystem is perpetually accumulating woody material, and then sharing and distributing the effects of its decomposition. As woody biomass decomposes, it releases carbon dioxide and cycles nutrients back into the soil, supporting trees' own progeny and a multitude of other species.

In the forest, we can understand decay as an essential process of ecosystem cycling, but when it comes to lumber, decay is a crisis. Wood's natural biodegradability is a problem to fix in an otherwise ideal building material. As a strong, lightweight, renewable, and carbon-sequestering material, wood is a poster child for sustainable construction,[6] but its very ability to biodegrade is problematic from other perspectives. A whole industry of techniques, treatments, and practices has evolved to arrest wood's decay and to slow down wood's propensity to become soil again. Naturally decay-resistant woods are prized for their durability in exterior constructions, for their ability to delay their own decomposition. This chapter focuses on one such wood: ipe (pronounced "ee-pay"), one of the most popular tropical hardwoods used in North America and installed in the first phase of the High Line, a park established in 2009 on the former elevated rail line of the West Side Improvement (discussed in Chapter 3). The case of ipe on the High Line exposes the controversies and implications of the trade in highly valued tropical hardwoods, as well as concerns over durability and obsolescence in construction materials, all within a context of rapidly changing post-industrial landscapes and the outsourcing of industry.

Always Living

The genus name of the coastal redwood, *Sequoia sempervirens*, is usually translated as "evergreen," signaling how the conifer retains its foliage throughout the winter, but some choose to translate the Latin as "always living."[7] The latter translation is apt because beyond the seasonal retention of its needle-like leaves, the species is adapted to keep on living: it regenerates quickly by sprouting seeds after disturbances; it is low in resins and fire-resistant, and the sapwood is particularly rich in polyphenols, making it impenetrable to fungi and insects. So well-equipped to respond to stresses that might wipe other species out, coastal redwoods thrive as some of the largest and oldest organisms in the world, towering more than 300 feet above the Pacific coast of California, some for 2,000 years at a time.[8]

Awe of these trees was matched by desire for their valuable lumber. Old-growth redwoods yield durable, beautiful, warm-toned lumber with a straight grain – and in mammoth volumes. Early settler loggers struggled to maneuver such massive logs out of the forest with oxen, and then cut them, so they sometimes used dynamite to break the logs into manageable pieces. But westward migration, logging, sawing, and transportation technology advanced, sending this unique and abundant high-quality lumber around the world for buildings, stadium seats, chemical vats, farm structures, fine carpentry, and highway

Figure 5.3 **Redwood logs on the way from forest to mill, between 1909 and 1920**
Source: Library of Congress Prints and Photographs Division, LC-F81–3364 [P&P].

bridges;[9] whole northern coastal cities were constructed out of adjacent redwood groves. A single mature redwood, a 1937 report by the U.S. Department of Commerce claimed, could provide enough lumber to build twenty-two new homes, and its bark – sometimes 2 feet thick – could be used for insulation.[10] The report argued that these trees were so plentiful, and the foresters' desire to maintain the resource so great, that the forests could be sustainably managed (balancing growth and removals) without any cutting restrictions at all.[11]

As this surge of redwood found its way into many types of construction, it also became a new element of modern domestic life: the residential deck. In the post-war years, as suburban residential construction boomed, the private back yard and residential wooden decks became symbols of the American dream. The deck bridged interior and exterior spaces, introduced a new kind of outdoor leisure, and could complement modern architecture's lines. Modern landscape architects like Garrett Eckbo, Robert Royston, and Thomas Church experimented with new forms of domestic outdoor landscaping, incorporating non-traditional materials and forms into the suburban garden in California.[12] Church's Martin Beach House design in northern California featured a redwood checkerboard deck and zig-zagging benches, and for the iconic Donnell Garden in Sonoma County, Church designed wooden decks perforated with existing trees, creating a surface that seemed to float.[13] Wooden surfaces were used to unite the inside and outside, as a new surface on which to fashion a modern life. Church himself even acted as a spokesman for the California

Figure 5.4 Landscape architect Thomas Church featured in advertisement for the California Redwood Association, 1956

Source: Originally published in House and Home Magazine, April 1956.

Redwood Association, and in a 1956 ad, he called redwood one of his "most versatile materials."[14]

While California redwood populations dwindled, the wood was still marketed as if its supply were infinite. "This John Chipman bench was planted 500 years before Columbus sailed for America," reads a furniture advertisement from a 1973 issue of *Landscape Architecture Magazine*.[15] The sturdy slatted bench is shown sitting atop a colossal redwood stump, surrounded by a lush forest canopy. "When you have a site furnishing job to do, think about Chipman in 1,000 year-old Redwood," the ad says. "Even if your benches only have to last another 100 years." By the 1980s, landscape architects' embrace of old-growth redwood had waned. Harvest rates plummeted because of the near-decimation of tree populations, and many of the remaining stands had been incorporated into parks and preserves. As access to old-growth lumber declined, other decay-resistant wood products, primarily chemically treated softwoods and tropical hardwoods, entered the market.

Because people believed that the United States had such vast supplies of cheap lumber, wood preservation developed more slowly in the United States than it had in Europe. But the combination of rising wood prices, depletion of high-quality local wood stocks, and increasing material performance requirements from the construction industry stimulated the American wood-preservation industry.[16] By sucking sap and air out of wood pieces in a vacuumed chamber and then injecting preservatives (such as Creosote, which dominated the industry in the early twentieth century) into its tissues, wood preservation offered a functionally different material, one that could surpass the decay-resistance of the best tree species. With the establishment of the American Wood Preservation Association, and the U.S. Forest Products Laboratory conducting and disseminating research as well as formalizing testing protocols and standards, a new material market was born. Wood preservation was pitched as a means to address the nation's rapidly depleting forest cover, as preservation would not only elongate the service time of wood, but would also allow a greater variety of species to be used. Mark Aldrich argues, however, that these aims weren't entirely realized because preservatives were often used on already durable woods like southern pine and fir.[17] "Preservation preserved markets," Aldrich concludes, "not forests."[18]

By the last quarter of the twentieth century, preserved or pressure-treated (PT) wood was a familiar product. While the oil-borne creosote and pentachlorophenol were widely used for railroad ties, utility poles, and industrial applications, they weren't used indoors or in more public settings because of known toxicity risks. But PT wood, the green-tinged lumber, became a familiar mass consumer good – affordable, reliable, and ubiquitous in playground structures and suburban decks. Chromated copper arsenic (CCA), a waterborne preservative under pressure, became the leading chemical of the industry.

Figure 5.5 Interior view of wood preservation cylinder, with incoming load

Source: Figure 14–3, in Forest Products Laboratory, *Wood Handbook – Wood as an Engineering Material. Gen. Tech. Rep. FPL – GTR – 113*. Madison, WI: U.S. Department of Agriculture, Forest Service, Forest Products Laboratory.

However, while it combatted fungi-rot and bonded tightly with wood, it also leached its highly toxic compounds into soil and water.[19] As well, the recycling of PT lumbers is problematic, and treated wood must be disposed of as a toxic material. Once treated, wood can no longer do what it otherwise does best: decay. In 2003, the Environmental Protection Agency called for a voluntary ban on using lumber treated with CCA.[20] As pressure-treated lumber began to be associated with serious human health risks, some CCA-treated playground structures were dismantled and sent to contained disposal facilities.

As redwood populations disappeared and conventional PT lumbers raised concerns, another source of durable woods surfaced: tropical hardwoods. While Americans specified tropical woods such as ekki (*Lophira Alata*) from West Africa and greenheart (*Chlorocardium rodiei*) from Guyana for railroads and public works as early as the 1940s, they privileged the cheaper CCA-treated southern pine.[21] In the second half of the century, U.S. tropical lumber imports increased as local hardwood prices rose, markets opened up, and political incentives towards exports were established in countries like Brazil.[22] Tropical lumber imports to the United States increased by 400 percent between 1960 and 1980.[23] And by the 1980s and 1990s, the advertising pages of *Landscape Architecture Magazine* offer benches in teak from Burma, as well as ipe, mahogany, and purple heart from South America.[24] North American designers were gaining newfound access to previously unknown woods – all promising better-than-redwood properties and without toxic chemical treatments.

When the iconic redwood decking of Thomas Church's Fay Garden in San Francisco was restored in 2006, it was replaced with ipe,[25] a tropical hardwood sourced primarily from Brazil and Peru. And when the first phase of the High Line was constructed on the obsolete elevated rail line of the West Side Improvement, ipe was the wood of choice for its decks, stepped seating, and furniture. Ipe's exceptional decay-resistance makes it one of the most popular types of commercially available tropical hardwood for higher-end residential decks, public parks, and boardwalks. Like redwood, ipe lumber resists rot and is dimensionally stable; unlike redwood, it is extremely hard and dense – nearly five times as heavy as old-growth redwood.

And like old-growth redwood, it is slow-growing, so its use is controversial, raising questions about the sustainability of its harvest. Four months after the High Line's opening in 2009, activists involved with the Rainforests of New York campaign hung a banner on the elevated park's Tenth Avenue Square,

Figure 5.6 **Rainforests of New York Action on Phase 1 of the High Line, 2009**

Source: Photograph by Rainforest Relief.

which is made with ipe certified by the Forest Stewardship Council (FSC). The banner read, "High Crime on the High Line: FSC Lies, Amazon wood is not sustainable." The campaign, initiated by members of Rainforest Relief and New York Climate Action, opposes the use of tropical hardwoods such as ipe in public landscapes, claiming that New York City – with the nation's largest park system and one of its most extensive boardwalk networks – is the largest municipal consumer of tropical hardwoods in North America.[26] When Rainforest Relief began speaking out about this issue, it promoted the use of FSC-certified tropical lumber, explains Tim Keating, the group's executive director.[27] But since then, the group has grown skeptical that certification is a reliable indicator of sustainable forestry practices when old-growth forests are involved.

Chains, Routes, Networks

Belém is located at the northern tip of Brazil, where the Pará River, intermingling with the southern arm of the Amazon River, drains into the Atlantic Ocean. The river is also the route by which forest products have traveled to Belém, the primary port of trade for Amazonian goods since Portuguese occupation. Wealth came in pulses to colonial powers with the extraction of valuable forest goods, largely through the enslaved labor of indigenous and African people: first cacao and then rubber.[28] Today, the city of Belém is still the gateway for Amazonian forest products and home to major environmental and research organizations as well as lumber exporters, including Timber Holdings USA – the lumber supplier for the High Line.

The Instituto Floresta Tropical (IFT), also based in Belém, straddles these domains of environmental protection and forest extraction.[29] The group conducts independent research and provides training and certification in sustainable forestry for logging companies, government agencies, and community landholders. When I asked Marco Lentini, the then technical manager of IFT, if I could see some ipe growing in its natural habitat, he invited me to visit IFT's Roberto Bauch Forest Management Center in Paragominas, a day's drive southeast of Belém. But he warned that we might not see many live trees, saying "looking for ipe is like following a trail of stumps."[30]

On the road to the Paragominas, Marlei Nogueira, who teaches sustainable forestry courses for IFT, points out different scenes of land conversion: freshly cleared forest tracts, fields of domed charcoal ovens, sawmills, and trucks laden with massive logs.[31] As we get closer to the camp, the forest gets denser, and we pass through a series of gates marked with the logos of IFT, other non-governmental and governmental organizations, and Cikal – one of the largest wood exporters in the state of Pará, owner of the land, and a partner in the center. The camp buildings seem minuscule compared to the enormous forest that surrounds it. A larger structure houses teaching areas where course participants meet before going out to the demonstration forests for training. Simple roofed structures are strung with hammocks where forest trainees and Cikal workers

Figure 5.7 **Logging truck, Paragominas, Brazil, 2012**

sleep. And beyond the central camp area, a soccer field is the only clearing in an otherwise endless forest. At night, as I listen from my hammock, the forest blasts from all directions. During the day, huge trees and scrambling vines dominate the visual landscape, but in the darkness, everyone else makes their presence known. All of those species – living in, on, and between trees – add their sounds to a sublime and deafening drone.

The next morning Marlei shows us the management plan maps that they use to harvest. The 5,000-hectare site is divided into 100-hectare work units. Four of these are harvested each year, and then they rest for thirty more years. The maps show which trees will be cut, in which direction the workers will haul them, and then where the trees will be left to be picked up by larger trucks. The courses that IFT teaches cover the basics of low-impact felling, skidding, and log storage, as well as how and where to build roads, and how to sequence this activity in relation to the time it takes to regenerate. Harvesting is as much about extracting wood, Lentini explains, as it is about making particular light gaps to facilitate the growth of particular species.

Marlei clears a path through the vines and the dense undergrowth towards a single mature purple ipe. At 1.5 meters in diameter and 40 meters high, the tree dwarfs those around it. The species, *Tabebuia impetiginosa*, is locally known as *ipe roxo* for the flush of purple flowers that appears after the dry season – a trait that makes it a popular urban street tree in Brazil. *T. impetignosa* and *T. serratifolia* (known as yellow or *ipe amarelo*) are two of the most exploited in the group of species collectively marketed as "ipe." These two species range from the Brazilian Amazon to Mexico, yet they appear at low densities; a mature tree is found just once every 3 to 10 hectares.[32] Their

Figure 5.8 **Mature ipe tree, Roberto Bauch Forest Management Center (Instituto Floresta Tropical), Paragominas, Brazil, 2012**

slight, winged seeds get caught in foliage as they fall from the canopy and are quickly consumed by tapir and deer on the forest floor. The seedlings, once established, require specific lighting conditions controlled by the size of forest gaps. Lentini explains that these traits, along with characteristic slow growth (less than 0.5 to 0.7 cm growth in diameter at breast height per year), produce a population comprising mainly old, large adults and few juveniles.

Marlei makes a small incision behind the tree's bark and removes a slice of the sapwood for us to inspect. At his prompting, we try to rip the shiny, yellow tissue with our fingers and find it impossible. Ipe species have thick-walled fiber cells that produce extremely dense and strong wood. The Janka Hardness Test, the industry standard for measuring the hardness of wood, determines the pressure required to embed a small steel ball into the surface of a wood board. At 3,680 pounds of force, ipe's Janka score is three and a half times greater than teak, and nearly nine times that of second-growth redwood.[33] Ipe wood is so hard that the U.S. Navy considered it for producing ball bearings during World War II; it is so heavy that it sinks in water.[34] As the tree grows and sapwood is converted to heartwood, biochemicals known as extractives are deposited into heartwood cells; this occurs in many species, but the particular extractives that accumulate in ipe species make the wood especially resistant to decay. These biological qualities make ipe a highly valuable commodity. According to Brian Lotz of Timber Holdings USA, a cubic meter of finished ipe lumber products is worth some $3,000. At that price, and given the amount of wood that is

Figure 5.9 **Ipe sapwood, 2012**

wasted in the making of finished products, a single mature tree could be worth around $9,000.[35]

We spot a seedling at the base of the large tree, and Nogueira picks off one of its five-leaflet leaves, showing us how it matches the logo on his IFT uniform: a stylized ipe leaf. He explains that ipe was chosen as the institute's emblem because it represents a great conundrum for the region. The desire for highly valuable yet sparsely distributed trees like ipe exacerbates deforestation in the Amazon. Deforestation patterns typically start with the harvesting of valuable species, then clearing the remaining biomass for charcoal production, and finally planting crops for cattle grazing. The pursuit of high-value species has pushed the logging frontier further into unlogged forests, catalyzing road construction and agricultural production. And as desired species are eradicated, the logging frontier advances even faster.[36]

The harvest of big-leaf mahogany (*Swietenia macrophylla*), also prized as a decking material in the United States, is a textbook example of these dynamics. The species faced extinction due to illegal logging and a rampant export market, and so in 2002 it was listed in Appendix II of the UN-chartered Convention on International Trade in Endangered Species (CITES). This regulation still allows the wood to be traded, but it limits the quantity and sizes of trees that can be harvested. It has alleviated some of the pressure on mahogany populations. But this pressure is transferred to other species like ipe, whose

Figure 5.10 **Marlei Nogueira and the Instituto Floresta Tropical logo in the shape of an ipe leaf, 2013**
Source: Photograph by author.

export market increased by 500 percent between 1998 and 2004. As ecologist Marc Schulze concludes, ipe is "the new mahogany."[37]

Forest engineer Eduardo Eguchi explains how the Brazilian government regulates logging. Companies must produce a "sustainable forestry management plan" that includes an inventory of all commercial trees with a diameter at breast height larger than 40 centimeters and a description of species to be harvested. The Brazilian Institute of Environment and Renewable Natural Resources (IBAMA) is responsible for reviewing management plans, granting licenses, and conducting inspections. Within a licensed site, logging companies that use heavy machinery are limited to taking 30 cubic meters of lumber per hectare for each thirty-five-year cutting cycle. Along with the protection of any endangered species, 10 percent of any species in a lot must be retained as a seed source, no species with fewer than three individuals per 100 hectares may be harvested, and unless otherwise legislated, no trees smaller than 50 centimeters diameter at breast height (dbh) may be cut.[38]

It has been illegal to export whole logs from Brazil since 1969. The sawmills of the Icoraci District, to the north of Belém, process logs that arrive by river and road from various regions of the Amazon. Marcia Teixeira of Timber Holdings USA offered to take me to one of the mills that her company purchases from. On the day of our visit, the mill is cutting ipe almost exclusively. Mounds of behemoth logs surround the roofed, open-sided structure. Inside the mill building, activity circles around the main saw: a colossal hook anchored to the trusses is lowered to catch the edge of

Figure 5.11 Ipe logs and boards, sawmill, Belém, Brazil, 2012

Figure 5.12 Ipe log being milled, Belém, Brazil, 2012

one of the logs. The hook flips it so that it is properly aligned, as three men slowly guide it towards the blade. With a set of carefully calibrated passes, the massive log becomes dimensional lumber, perfect, straight, and anonymous.

At the front of the building are floor-to-ceiling stacks of smooth, warm-hued ipe 2×4s drying. Teixeira works with the mill's grader to check the quality and verify the moisture levels of the lumber. Boards are air- or kiln-dried to 12 percent humidity before shipping to ensure lighter and cheaper freight. The ipe that Timber Holdings USA buys comes primarily from the states of Pará and Mato Grosso. Logs are typically transported from interior forests to rivers by truck, and then by barge to port cities like Belém. Timber Holdings USA purchases sawn wood and ships it directly from Brazilian mills to New York City's port. And from there it is distributed to their warehouses in Wisconsin, New Hampshire, New Jersey, Texas, and California.[39]

In a hotel lobby in Belém, Teixeira walks me through the string of documents required to secure a legal shipment of lumber from Brazil to the United States. Each shipment of sawn lumber is associated with a unique number and set of documents that describe its chain of custody. In 2006, IBAMA instituted the Forest Origin Document system, whereby all forest products are tracked digitally in real time. According to the Brazilian Government, this system, along with new surveillance methods, has significantly increased its ability to control illegal logging and curtail deforestation. Teixeira's job involves the meticulous administration of Timber Holdings USA's lumber on its way out of Belém, maintaining records of the locations, handlers, and licenses associated with each link of the chain, from forest to importation. Tightened regulations, along with amendments to the U.S. Lacey Act, which prohibits the trade of any

Figure 5.13 **Ipe lumber stacked for shipment, Belém, Brazil, 2012**

illegally harvested plants or animals, have required companies to be even more diligent. Timber Holdings USA has also developed a continuing education course to help clients navigate the controls and write product specifications, Brian Lotz tells me. The paperwork is so arduous, Teixeira says, that "illegal companies can't survive."

On the road back from the IFT Field Station, we see officials from IBAMA and SEMA (Pará's Environmental Secretariat) stopping trucks loaded with logs to check their documents. This legal paperwork, Marlei says, is essential, but it can also provide an administrative mechanism for concealing illegal wood. In 2012, the BBC reported that some logging operations had bribed or bullied landowners to produce legal forest management plans, which were then used to cover illegal lumber from a different forest. The scam is known as "heating" the wood, and the vast terrain and complex network of actors related to logging in the Amazon make it challenging to control.[40] Illegal logging in the Brazilian Amazon decreased substantially in the 2000s, but a recent report by Chatham House estimates that government response to illegal logging has since weakened, and that between 36 and 78 percent of natural forest production in Brazil is due to illegal logging.[41] Just as it is nearly impossible to fully understand the extent of illegal logging, it is difficult to grasp the labor conditions that surround it. In some cases, illegal logging activities involve a miserable arrangement known as "debt slavery." Migrant laborers, assembled into work crews by armed intermediaries who supply their basic provisions at greatly inflated costs, find themselves indebted into servitude at the risk of their lives. In Pará alone, an estimated 25,000 people working in landscape conversion live in this abysmal condition.[42]

Even if all illegal logging could be stopped, that might not stabilize ipe populations. To actually produce a sustained yield of a given species, the cutting intervals and quantities must be calibrated to match regeneration rates. The Tropical Forest Foundation, an international not-for-profit, asserts that the harvests allowed under Brazilian law are too generous, and that conventional, legal logging in Brazil is detrimental to its sustainability. The foundation and others promote "reduced impact logging," which involves longer cutting cycles, harvesting methods that protect neighboring trees, measures to reduce soil erosion, and improved working conditions.

Amazonian deforestation became the symbol of globalized environmental calamity during the 1990s; a shift in the discourse saw deforestation no longer as a local indigenous land-management problem, but rather as a global concern.[43] Various non-governmental organizations, such as the World Wildlife Fund and Friends of the Earth, developed the Forest Stewardship Council (FSC) in 1993 – a non-state, market-driven evaluation program.[44] The FSC has created the most recognizable third-party certification system for sourcing wood in the United States and oversees third-party certifiers who administer two forms of certification: forest management and chain of custody. In return, companies may gain access to higher-price markets for certified products. Ian Hanna, former Director of Business Development of the FSC in the United States, estimates that approximately 5 percent of the American forest-product market (including everything from cardboard packaging to timber) is FSC certified.[45] This is comparable to the current consumption of certified organic food.

As of 2018, some 6.5 million hectares of Brazilian land are certified by the FSC.[46] The larger part of this reflects plantations, and the rest is "natural" forest, which includes a spectrum of forest types, from relatively old-growth forest to re-naturalized former pasture.[47] The FSC doesn't explicitly ban the harvest of old-growth forests, Hanna tells me, except where they are exceedingly rare. The FSC maintains that the careful management of all types of forests is the best way to protect them – unmanaged forests receive no protection, Hanna says. The FSC's forest-management certification is based on ten principles that address environmental impacts, the rights of indigenous peoples and workers, the maintenance of high conservation-value forests, and other issues. Principle Five requires that cutting rotations, extractable volumes, and the percentage of seed trees retained must be based on data on the various species biology. Hanna says that at the level of the forest, the sustained yield of wood is strictly met, more so than at the species level.

A 2008 study by Schulze, James Grogan, and Edson Vidal challenges the assumption that FSC-certification guarantees sustained yields.[48] They studied all FSC certificates issued in the Amazon from 1993 to 2005 and found that while the standards are comprehensive, assessments are administered subjectively, which results in inconsistent levels of compliance with regards to harvest operations, forest management, and monitoring. If companies in the Amazon were strictly held to all standards required for certification, the researchers conclude that none would currently qualify. The authors found no evidence of species-level management plans in their analysis at the time of their study.

While certified ipe does come from forests managed with current best practices, it is not, they argue, being managed to attain a sustained yield. They underscore the importance and major gains of certification, but think that lack of clarity about what standards are being met produces unreal impressions of its success – and undermines the process.

In another paper published the same year in *Biological Conservation*, the same authors, along with Lentini and Chris Uhl, analyzed the population dynamics of purple and yellow ipe (*T. impetignosa* and *T. serratifolia*) on sites harvested under conventional and reduced-impact logging regimes, and projected the impacts of future harvests.[49] Their analysis found that even with reduced logging intensity, it is not possible to sustain the production of ipe lumber under current "sustainable" harvest levels; the numbers of trees will continue to decline. The authors recommend further protection of the species, similar to those granted to mahogany. This would involve applying stricter regulations about how much of the species could be harvested from the forest, and incorporating silvicultural techniques that could improve chances of seedling survival.

In the experimental plots surrounding IFT's Forest Management Center, these techniques are being tested on plantations in forest gaps from seedlings

Figure 5.14 (A) Silviculture experimental plot, showing one-year-old growth, and (B) ten-year-old growth

Source: Roberto Bauch Forest Management Center (Instituto Floresta Tropical), Paragominas, Brazil, 2012.

collected from mature trees. Paricá (*Schizolobium amazonicum*), mahogany, and ipe are planted together to take advantage of their distinct cutting cycles; paricá will be harvested every thirty years, mahogany every sixty years, and ipe every ninety years. We visited two forest plantations, one planted the year before and another planted ten years earlier. The young plantation looks like a small garden of neat rows of seedlings that Marlei carefully tends to. At the ten-year-old plantation, I can hardly tell the planted trees apart from the other growth. Plantation timber, which grows faster than its forest-grown counterparts, tends to be inferior in quality. However, it is these techniques that might one day make the commercial harvest of slow-growing trees like *Tabubuia spp.* a more tenable proposition.

Sense and Obsolescence

As you walk south along the High Line from the northernmost end at the Hudson Yards towards the park's terminus at Gansevoort and Washington Street, you follow the rail path that live cattle and produce traveled as they arrived in New York City between 1934 and 1980. Lifting trains off of Eleventh Avenue – known as Death Avenue for its fatal mix of trains and pedestrians – was one of the primary agendas of the southernmost node of the 1930s West Side Improvement plan. Three storeys up, trains of the nicknamed "High Line" passed between store yards and buildings, entering and exiting warehouses, and casting moving shadows onto the streets below. Over the nineteenth century, the neighborhood had developed into a critical shipping and manufacturing hub, composed of large open lumber yards, steam saw mills, brass and iron works, and stone-cutting operations.[50] But as real estate property values rose, these land uses gave way to higher-density industrial, commercial, and residential buildings – and eventually the elevated rail line.

After forty-six years in operation, the High Line stood vacant for thirty more. By the late 1960s, rail traffic slowed as truck and highway transport gained prominence, and in 1980 all traffic on the High Line came to a halt.[51] As manufacturing and industrial uses shifted out of West Chelsea, the neighborhood became a space for artists, then galleries, and then more services, financial institutions, and high-end real estate development. And as segments of the heavy steel structure became impediments to new development, they were taken down, first below Bank Street in 1960, and then below Gansevoort Street in 1991.[52] Some property-owners, seeing the structure as unattractive, a magnet for illicit activity, and a threat to property values, lobbied to have it demolished entirely. In his last weeks in office, former Mayor Rudy Giuliani signed off on the High Line's demolition. But in the final hour, a movement to repurpose the structure by citizen-business consortium Friends of the High Line, city politicians, and the site's owner eventually prevailed; following a high-profile design competition, the first phase of the park opened in 2009.[53]

It is difficult to overstate how popular this project has been. The park is one of the city's most visited tourist destinations and a triumphant precedent for adaptive reuse of industrial structures, inspiring many similar projects in cities around the world. It is lauded for its win-win-win arrangement: merging the preservation of industrial structures, new types of public space, *and* economic growth.[54] Or as its design team envisioned: "a post-industrial instrument of leisure, life, and growth."[55] The High Line, as one sociologist put it, is an archetypal neoliberal space. It is emblematic of new landscape projects built under neoliberal urban agendas, which, through different mechanisms, shift public services towards the private sector, all taking place within the aesthetics of obsolescence. Many of the most-acclaimed North American landscape architecture projects of the last two decades are sited in more affluent cities, in the remnants of twentieth-century industry, whether rail yards, shipping ports, landfills, or manufacturing areas. Their transformation offers the return of industrial structures to the public realm, often accompanied by the remediation of contaminated soil. Like the High Line, these sites index the migration of industry out of urban centers (and often out of the country), their replacement with service and financial industries, and the displacement of the urban working class. In the 1870s, if Central Park was the antidote to the growing industrial city, and in the 1930s the renovation of Riverside Park was the means to negotiate with the industrial city in full swing, today parks occupy the obsolescent structures left behind after industry's disappearance.

Following new urban financial models, these new parks are often required to be self-sustaining, generating capital through commercial activities and corporate partners. Their existence requires partnering with non-profit conservancies, like Friends of the High Line or the Central Park Conservancy, which typically direct and raise private funds to support the park's programming, maintenance, and operations, which were previously assumed by city agencies. Corporate partners contribute as park stakeholders and tend to support higher-end consumption and passive activity – preferences enforced by heightened levels of security and managerial control.[56] On the one hand, parks like the High Line are of exceptional design quality, well-built with high-end materials (like ipe), and offer unique park experiences. But on the other hand, many of the park's most desirable qualities are made possible through measures of control that ultimately exclude. That visitors to the High Line are disproportionately white is just one significant example of exclusionary privilege in this type of post-industrial landscape.[57] While a visitor to the High Line might enjoy a late spring nap on a wooden chaise longue, feeling secure in the complex of guards, park attendants, and vendors, in other parts of the city, someone attempting to sleep in a park might be fearful of their personal security and police aggression.[58] The result is a very different type of park than exists in poorer or racialized neighborhoods, where less social capital exists for developing such management systems, policing is more aggressive, and park budgets dwindle.

According to the founders of the Friends of the High Line, James Corner Field Operations, Diller Scofidio & Renfro, and Piet Oudolf won the 2001 design competition in part because of how they engaged the structure's history of obsolescence.[59] The team saw value in the structure as a ruin, as a hidden space of marginal activity. They didn't propose new buildings on the line; instead they proposed to strip elements away to reveal what was there and develop a pathway system that foregrounds the relationship between the steel structure, new surface materials, and plants. Before the competition, Joel Sternfeld's photographs of the abandoned rail line popularized the sublime quality of pioneering vegetation – seeded over time by itinerant birds, trains, and other animals – growing through the gravel surface.[60] Similarly, the proposal offered what Corner describes as a furrowed landscape surface, an intermingling of plants and paving through the use of tapered pre-cast concrete planks.[61] And Piet Oudolf's planting scheme was seen as an idealized version of the vegetation that slowly colonized the site. By framing the industrial apparatus and plant growth over time, the designers sought a park that would be a slow, otherworldly, durational space.

Wood isn't a dominant material on the High Line, but it does perform a critical function in the project's material palette. Within the elevated steel structure, wood domesticates the industrial space, makes it warmer to the hand, more comfortable to the seat. At the Tenth Avenue Square (where the Rainforest Relief and New York Climate Action demonstration took place), an ipe deck, punctuated with trees and benches that rise out of the deck's surface, descends via steps and ramps right through the steel structure, offering you a window to the traffic below. From here you're literally within the structure, neither on top nor below, but insulated in the warmth of the wood. On another stretch between 14th and 15th Streets, the park opens up a broad vista onto the Hudson River. There, a line of generous wooden chaise lounges, constructed of sturdy 2×6 ipe boards, line up like rail cars on repositioned steel tracks.[62] Visitors stretch out facing the afternoon sun, sometimes dozing off in the sunshine, nestled amongst plantings. Elsewhere ipe benches "peel up" as extensions of the concrete paving units below, another iteration of the idea of intermingling surfaces.[63] In the later phases wood bleacher seating recalls piles of stacked lumber and allows for more open-ended seating, and an interminable curving bench signals the momentum of the former train path.

Figure 5.15 **Tenth Avenue Square, the High Line**
Source: Photograph by Cynthia Goodson, 2013.

Figure 5.16 **Lounge seating, the High Line**
Source: Photograph by Cynthia Goodson, 2013.

Figure 5.17 **Reclaimed teak benches in Phase 2 of the High Line, New York City, 2012**

In these instances, the wood surface mediates between the park visitor and the industrial structure; it is the material that one most often touches with skin, the material that even invites you to lie down. It seems "natural" and clean; in other post-industrial landscapes, wood is often the boardwalk that hovers above, separating people from a sensitive ecosystem or toxic ground. The High Line design team's decision to use FSC-certified ipe in phase one of the High Line aligned with current best practice for high-traffic public landscapes that had the budget. Their specification of an expensive but durable wood would minimize maintenance and delay replacement, and the FSC certification is by all conventional standards a respectable bar for sustainability considerations. And for such a high-profile project, its success would depend on the durability of its details.

The 2009 protest of ipe use on the High Line by members of Rainforest Relief and New York Climate Action was part of three decades of activism against New York City's consumption of tropical hardwoods. Members of the groups have dropped banners from the top of the Coney Island Parachute

Jump in 1998, the flagpoles at City Hall in 2010, and the arch of Washington Square Park in 2011.[64] The two groups joined to form the Rainforests of New York campaign, and as the name suggests, their message is about bringing this distant wood source – the rainforest – to the awareness of New Yorkers so that the implications of its use can be confronted. Over decades, members of the groups have led tours and made videos that identify the species and origins of the city's wood, highlighting ecological and human-rights issues and abuses. They have met frequently with city and state agencies, community groups, and architects; organized resistance campaigns against new construction projects;[65] and advocated for tropical hardwood alternatives. The High Line, admired for its identity as ultimately local – with its reference to the neighborhood's industrial past and its vegetation pioneers – was a strategic location from which to invoke the very distant and abstract consequences of old-growth tropical hardwood use.

When Mayor Michael Bloomberg announced to the United Nations General Assembly in late 2007 that he would reduce New York's tropical hardwood use by 20 percent,[66] Rainforest Relief's Tim Keating was surprised: "This is the first time that anyone ever stepped up and agreed with us. And he took the steps to get the agencies to get them to pay attention to these issues." A report that followed Bloomberg's announcement laid out a plan for reducing tropical hardwood use by substituting recycled plastic lumber, domestic and non-tropical hardwoods and softwoods, concrete, as well as certified sustainable tropical hardwoods.[67] Domestic black locust (*Robinia pseudoacacia*) is particularly promising. It is a species native to the United States, compares in strength and hardness to ipe, is harder than teak and white oak, and yet it is fast growing. The species was the focus of a 2011 American Society of Landscape Architects presentation by the landscape architect Michael Van Valkenburgh of MVVA and Don Lavender of Landscape Forms, which recently introduced black locust to its palette for site furnishings. In MVVA's Brooklyn Bridge Park, for example, black locust benches line Pier One promenade, and picnic tables

Figure 5.18 (A) Black locust benches, Pier One Promenade, (B) Reclaimed long leaf yellow pine picnic benches and tables, Brooklyn Bridge Park, Michael Van Valkenburgh Associates, Inc.

Source: Photographs by Elizabeth Felicella, 2011.

and benches are made of reclaimed long leaf yellow pine. Other decay-resistant materials have also proliferated in the past decade, including heat-treated wood, which physically alters wood cells to prevent fungal penetration, as well as wood treated with silica and other non-arsenic-based chemical treatments.

But substitution is not always so easy, the report describes. Construction assemblies are calibrated to materials of particular strength, and so, for example, replacing the unusually strong tropical greenheart (*Chlorocardium rodiei*) on the Brooklyn Bridge Promenade with weaker materials would require an overhaul of the support structure. Furthermore, a section of New York State Municipal Law makes it illegal to privilege FSC-certified lumber, as it is considered physically equivalent to non-certified lumber.[68] And even when FSC-certified ipe might be sought, it can be difficult to find. At the time of my visit to Belém, two of Timber Holdings USA's lumber suppliers had access to FSC lumber, but neither had any certified ipe. Teixeira predicted that there may be a small amount of FSC-certified ipe available a few months later, but it is difficult to count on in an otherwise vast market. FSC's Hanna sees this limited supply of FSC-certified ipe as an indication of the careful management of an uncommon species. He believes that ipe is overused because it is specified by convention, not function. When designers rely on certain popular woods like ipe, many other similarly performing woods are overlooked and undervalued (and with them, the land they occupy). The specification of less common species, says Amy Smith of the World Wildlife Fund (WWF), has the potential to shift value to the ecosystem rather than the single species and encourage the long-term management of the standing forest. The WWF, a longtime partner of the FSC, launched a set of tools for designers to navigate the dense literature on wood selection and find lesser-known timber species.[69] She cites angelim vermelho (*Dinizia excelsa*) and curupay (*Anadenanthera colubrina*) as two promising tropical hardwood alternatives to ipe.[70]

Since Bloomberg's declaration, the NYC Department of Parks and Recreation has stopped specifying tropical wood in park benches entirely. Ruby Wei, Director of Specifications for the Parks Department, tells me that today, most new benches are made in recycled plastic lumber, and sometimes in steel. They specify wood – primarily white oak or thermally modified lumber – only where there is a historical significance and they need to match existing benches in a site. They would like to use black locust, but it has been difficult to procure.[71] New York joins the company of San Francisco, Santa Monica, and Baltimore, cities that banned the use of tropical hardwoods for municipal projects in the 1990s.[72] The second phase of the High Line was finished with oil-treated teak, salvaged from Indonesian industrial and agricultural structures, instead of ipe. The third phase is now rich with reclaimed Angelique (*Dicorynia guianensis*), a deep amber tropical hardwood reclaimed from the Coney Island Boardwalk.[73] Reclamation is costly, as it requires milling and finishing to specifications, but as Tim Doody of Rainforest Relief pointed out, it reinforces some of the key ambitions of the project itself: "The reclaimed teak seems very much in keeping with the spirit of a park designed on an elevated train track that almost got torn down."[74]

Ipe earned its place in the landscape because it can withstand decades of tough weather, but Superstorm Sandy revealed that no wood is a match for a 13-foot storm surge. Striking post-storm photographs showing splintered piles of lumber, the remains of the famous Rockaways boardwalk, reveal how easily that which seems durable can be crushed in the face of climate-driven weather events. A *New York Times* journalist described one such pile as, "a hardwood hodgepodge of Angelique, teak, pine, ipe, Cumaru, and greenheart."[75] The crumpled boardwalks triggered debates about whether wood replacements will be able to withstand the more frequent storms resulting from climate change. In 2017, the boardwalk reconstruction was completed – in concrete.[76] Other boardwalks ravaged by the storm, like those in Avon, New Jersey, were rebuilt in ipe, despite a powerful campaign from Rainforest Relief and the New Jersey Sierra Club.[77] Tim Keating sees the devastation of Superstorm Sandy as a wake-up call and sees a direct connection between the destroyed boardwalks and the status of the forests that their wood comes from. "Sandy wiped out boardwalks from Maryland to New York," he says, "the fact that towns are willing to go back to tropical hardwood is incredible. . . . Materials are not a free ride – they are taken from somewhere, and we forget that at our own peril."

Seeing a piece of wood decay over time, you might first notice the softening of the surface, the slow digestion of the outer fibers, and then later, the wood becoming spongy or brittle before it eventually surrenders and falls apart. A sign of failure from some perspectives, this ability to degrade is also one of wood's most significant traits, unique in the palette of conventional construction materials. The sapwood cells have their own mechanisms of defense, but

Figure 5.19 **Rockaway Beach boardwalk following Hurricane Sandy, 2012**
Source: Photograph by Roman Iakoubtchik.

ultimately they give way, and the cycling of nutrients from their own demise ends up supporting the forest as a whole, cycling carbon and other nutrients through the soil. Their fragility is their future strength. Observing decay is a reminder of the complex relationships that allow ecological systems to thrive, endure, and regrow. Decay is a useful metaphor for thinking about how and what to construct in this age – we can think about the prior life and design for the afterlife of the materials that we engage with. At the same time, the process of decaying wood is also not a metaphor – but simply one of the pathways that sustains human life, and other life on which we depend.

Notes

1 Carol A. Clausen, "Biodeterioration of Wood," in *Wood Handbook: Wood as an Engineering Material* (Madison: US Department of Agriculture, Forest Service, Forest Products Laboratory, 2010), 3–4. Decay-producing fungi is the primary source of decay for trees in North America, but insects, molds, and stains all also transform and degrade wood in different ways.

2 Robert A. Zabel, *Wood Microbiology: Decay and Its Prevention* (San Diego: Academic Press, 1992), 168.

3 Ibid.

4 Walter C. Shortle and Kenneth R. Dudzik, "Wood Decay in Living and Dead Trees: A Pictorial Overview," United States Department of Agriculture, General Technical Report NRS-97 (Pennsylvania: U.S. Forest Service, 2012), 9–10; and Kevin T. Smith, "Compartmentalization Today," *Aboricultural Journal* 29 (2006): 174.

5 Clausen, "Biodeterioration of Wood," 4.

6 Daniel Ibanez, Jane Hutton, and Kiel Moe, eds., *Wood Urbanism: From the Molecular to the Territorial* (Barcelona: Actar, 2018).

7 Forest Products Division, US Department of Commerce, *California Redwood and its Uses* Trade Promotion Series No. 171 (Washington, DC: US Government Printing Office, 1937), 1.

8 Ibid.

9 Ibid., 2–3.

10 Ibid., 1.

11 Ibid., 5.

12 Marc Treib and Dorothée Imbert, *Garrett Eckbo: Modern Landscapes for Living* (Berkeley: University of California Press, 2005), 55.

13 Marc Treib, "Axioms for a Modern Landscape," in *Modern Landscape Architecture: A Critical Review*, ed. Marc Treib (Cambridge, MA: The MIT Press, 1994), 64.

14 Diane Harris, "Writing a Modern Landscape: Thomas Church as Author," *House and Home*, April 1956, 227.

15 Landscape Forms (Advertisement), *Landscape Architecture Magazine* 64 (1973).

16 George M. Hunt and George A. Garratt, *Wood Preservation* (New York: McGraw-Hill Book Company Inc., 1953), 3; Mark Aldrich, "From Forest Conservation to Market Preservation: Invention and Diffusion of Wood-Preserving Technology, 1880–1939," *Technology and Culture* 47, no. 2 (April 2006): 312.

17 Ibid., 331.

18 Ibid., 340.

19 J. William Thompson and Kim Sorvig, *Sustainable Landscape Construction* (Washington, DC: Island Press, 2000), 222.

20 U.S. Consumer Product Safety Commission, "What You Should Know about CCA-Treated Wood," *CPSC Publication 270*, www.cpsc.gov/PageFiles/122137/270.pdf.

21 "Another Wood for Boardwalks," *New York Times*, July 10, 1994, NJ6.

22 William C. Siegel and Clark Row, "US Hardwood Imports Grow as World Supplies Expand," *Forest Service Research Paper SO-17* (U.S. Department of Agriculture, 1965), 4; John O. Browder, "Brazil's Export Promotion Policy (1980–1984): Impacts on the Amazon's Industrial Wood Sector," *The Journal of Developing Areas* 21, no. 3 (April 1987): 286.

23 Martin Chudnoff, *Tropical Timbers of the World* (Madison: US Department of Agriculture, Forest Service, Forest Products Laboratory, 1980), 1.

24 See advertisements in *Landscape Architecture Magazine* 30 (July 1985): 541, 556; *Landscape Architecture Magazine* 80 (July 1990): 23.

25 Daniel Jost, "Changing Places: Degrees of Preservation: Three Residential Landscapes by Thomas Church Face Challenges Under New Ownership," *Landscape Architecture Magazine* (January 2009): 30–40.

26 Colin Moynihan, "Destination New York City," *Metropolis* 19, no. 8 (June 2000): 50.

27 Tim Keating, Personal correspondence, February 26, 2013.

28 Timothy Walker, "Slave Labor and Chocolate in Brazil: The Culture of Cacao Plantations in Amazonia and Bahia (17th–19th Centuries)," *Food and Foodways* 15, no. 1–2 (June 6, 2007): 75–106.

29 Instituto Floresta Tropical, http://ift.org.br/en.

30 Marco Lentini, Personal correspondence, May 28, 2012.

31 Marlei Nogueira, Personal correspondence, May 29, 2012.

32 Mark Schulze, James Grogan, Chris Uhl, Marco Lentini, and Edson Vidal, "Evaluating Ipê (*Tabebuia, Bignoniaceae*) Logging in Amazonia: Sustainable Management or Catalyst for Forest Degradation?" *Biological Conservation* 141, no. 8 (August 1, 2008): 2077.

33 David E. Kretschmann, "Mechanical Properties of Wood," in *Wood Handbook: Wood as an Engineering Material* (Madison: US Department of Agriculture, Forest Service, Forest Products Laboratory, 2010), 5–23.

34 Sarah A. Laird and Alan Robert Pierce, *Tapping the Green Market: Certification and Management of Non-Timber Forest Products* (London: Earthscan, 2002), 86.

35 Estimated based on 1.8–2.9 cubic meters of final products per tree. Marco Lentini, personal correspondence with author, May 28, 2012.

36 Helmut J. Geist and Eric F. Lambin, "Proximate Causes and Underlying Driving Forces of Tropical Deforestation," *BioScience* 52, no. 2 (February 2002): 150.

37 Schulze et al., "Evaluating Ipê," 2072.

38 Eduardo Eguchi, Personal correspondence, May 28, 2012.

39 Marcia Teixeira, Personal correspondence, May 28, 2012.

40 Sue Branford, "How Illegal Amazon Logging Persists," *BBC News*, November 29, 2012, www.bbc.com/news/world-latin-america-20408238.

41 Laura Wellesley, "Illegal Logging and Related Trade: The Response in Brazil," *A Chatham House Assessment*, 2014, 24, www.chathamhouse.org/sites/default/files/publications/research/20141029IllegalLoggingBrazilWellesleyFinal.pdf.

42 Philip M. Fearnside, "The Roles and Movements of Actors in the Deforestation of Brazilian Amazonia," *Ecology and Society* 13, no. 1 (2008): 29.

43 Andrea Zhouri, "Global-Local Amazon Politics: Conflicting Paradigms in the Rainforest Campaign," *Theory, Culture & Society* 21 no. 2 (2004): 77.

44 Benjamin Cashore, Graeme Auld, Steven Bernstein, and Constance McDermott, "Can Non-State Governance 'Ratchet Up' Global Environmental Standards? Lessons from the Forest Sector," *Review of European, Comparative & International Environmental Law* 16, no. 2 (July 2007): 159.

45 Ian Hanna, Personal correspondence, March 12, 2013.

46 Forest Stewardship Council International, "Facts and Figures," https://ic.fsc.org/en/facts-and-figures.

47 Vanessa Maria Basso, Laércio Antônio Gonçalves Jacovine, Aurea Maria Brandi Nardelli, Ricardo Ribeiro Alves, Edson Vidal da Silva, Márcio Lopes da Silva, and

Bruno Geike de Andrade, "FSC Forest Management Certification in the Americas," *International Forestry Review* 20, no. 1 (2018): 36.

48 Mark Schulze, James Grogan, and Edson Vidal, "Forest Certification in Amazonia: Standards Matter," *Oryx* 42, no. 2 (April 1, 2008): 229–39.

49 Schulze et al., "Evaluating Ipê," 2083.

50 Christopher D. Brazee and Jennifer L. Most, *West Chelsea Historic District Designation Report* (New York: New York City Landmarks Preservation Commission, 2008), 13.

51 Joshua David and Robert Hammond, *High Line: The Inside Story of New York City's Park in the Sky* (New York: FSG Originals, 2011), ix–x.

52 Ibid.

53 David W. Dunlap, "On West Side, Rail Plan is Up and Walking," *New York Times*, December 22, 2002, 43.

54 Kevin Loughran, "Parks for Profit: The High Line, Growth Machines, and the Uneven Development of Urban Public Spaces," *City & Community* 13, no. 1 (March 2014): 56.

55 The High Line, www.thehighline.org/design/design-team-selection/field-operations-diller-scofidio-renfro.

56 Loughran, "Parks for Profit," 62.

57 Ibid.; Alexander J. Reichl, "The High Line and the Ideal of Democratic Public Space," *Urban Geography* 37, no. 6 (2016): 919.

58 Loughran, "Parks for Profit," 62.

59 David and Hammond, *High Line*, 75, 77.

60 Joel Sternfeld, *Joel Sternfeld: Walking the High Line* (Göttingen: Steidl, 2012).

61 James Corner and Ricardo Scofidio, *Designing the High Line: Gansevoort Street to 30th Street* (New York: Friends of the High Line, 2008), 118–19.

62 Ibid., 64.

63 Ibid.

64 "A Protest at Great Height," *New York Times*, August 17, 1988, B3; Rainforests of New York, http://rainforestsofnewyork.net.

65 Jill P. Capuzzo, "Rainforest Politics Strides onto the Boardwalk," *New York Times*, June 24, 2007, www.nytimes.com/2007/06/24/nyregion/nyregionspecial2/24mainnj.html.

66 Sewell Chan, "Bloomberg Urges U.N. to Act on Climate Change," *New York Times*, February 11, 2008, https://cityroom.blogs.nytimes.com/2008/02/11/bloomberg-urges-un-to-act-on-climate-change; NYC Press Release PR-045-08, "Mayor Announces Plan to Reduce the Use of Tropical Hardwoods," February 11, 2008.

67 Rohit T. Aggarwala, "Tropical Hardwood Reduction Plan," *Memorandum to Mayor Michael R. Bloomberg*, 7–9, February 11, 2008.

68 Ibid., 9–10.

69 World Wildlife Fund, "Guide to Lesser Known Tropical Timber Species," www.worldwildlife.org/publications/guide-to-lesser-known-tropical-timber-species.

70 Amy Smith, Personal correspondence, March 22, 2013.

71 Ruby Wei, Personal correspondence, March 5, 2019.

72 San Francisco Environment, https://sfenvironment.org/policy/chapter-8-tropical-hardwood-and-virgin-redwood -ban; Larry Adams, "Baltimore Joins Timber Ban Bandwagon," *The Free Library* (March 1991), www.thefreelibrary.com/Baltimore+joins+timber+ban+bandwagon.-a010689306.

73 Alan Soloman, "It's Tropical on the Highline," *Sawkill Lumber Co*, February 13, 2015, http://sawkil.com/its-tropical-on-the-highline.

74 Rainforests of New York, "Amazon Wood Ejected from the Highline," http://rainforestsofnewyork.net/amazonwoodejectedfromhighline.

75 Liz Robbins, "Clash Over Future of Wood from the Storm-Torn Rockaway Boardwalk," *New York Times*, November 15, 2012, www.nytimes.com/2012/11/16/nyregion/rivals-vie-for-wood-from-storm-torn-rockaway-boardwalk.html?_r=0.

76 Luis Ferré-Sadurní, "Could the Rockaways Survive Another Sandy?" *New York Times*, July 13, 2017, www.nytimes.com/2017/07/13/nyregion/rockaways-beaches-hurricane-sandy.html.

77 Andrea Appleton, "Post-Sandy, Green Groups at Loggerheads with Plans to Rebuild Jersey's Boardwalk Empire," *Grist*, March 20, 2013, https://grist.org/climate-energy/post-sandy-rebuilding-jerseys-boardwalk-empire-puts-green-groups-at-loggerheads.

Bibliography

Adams, Larry. "Baltimore Joins Timber Ban Bandwagon." *The Free Library*, March 1991, www.thefreelibrary.com/Baltimore+joins+timber+ban+bandwagon.-a010689306.

Aggarwala, Rohit T. "Tropical Hardwood Reduction Plan." *Memorandum to Mayor Michael R. Bloomberg*, February 11, 2008, www.nyc.gov/html/om/pdf/tropical_hardwoods_report.pdf.

Aldrich, Mark. "From Forest Conservation to Market Preservation: Invention and Diffusion of Wood-Preserving Technology 1880–1939." *Technology and Culture* 47, no. 2 (April 2006): 311–40.

"Another Wood for Boardwalks." *New York Times*, July 10, 1994.

Appleton, Andrea. "Post-Sandy, Green Groups at Loggerheads With Plans to Rebuild Jersey's Boardwalk Empire." *Grist*, March 20, 2013, https://grist.org/climate-energy/post-sandy-rebuilding-jerseys-boardwalk-empire-puts-green-groups-at-loggerheads/.

Basso, Vanessa Maria, Laércio Antônio Gonçalves Jacovine, Aurea Maria Brandi Nardelli, Ricardo Ribeiro Alves, Edson Vidal da Silva, Márcio Lopes da Silva, and Bruno Geike de Andrade. "FSC Forest Management Certification in the Americas." *International Forestry Review* 20, no. 1 (2018): 31–43.

Branford, Sue. "How Illegal Amazon Logging Persists." *BBC News*, November 29, 2012, www.bbc.com/news/world-latin-america-20408238.

Brazee, Christopher D., and Jennifer L. Most. *West Chelsea Historic District Designation Report*. New York: New York City Landmarks Preservation Commission, 2008.

Browder, John O. "Brazil's Export Promotion Policy (1980–1984): Impacts on the Amazon's Industrial Wood Sector." *The Journal of Developing Areas* 21, no. 3 (April 1987).

Capuzzo, Jill P. "Rainforest Politics Strides onto the Boardwalk." *New York Times*, June 24, 2007, www.nytimes.com/2007/06/24/nyregion/nyregionspecial2/24mainnj.html.

Cashore, Benjamin, Graeme Auld, Steven Bernstein, and Constance McDermott. "Can Non-State Governance 'Ratchet Up' Global Environmental Standards? Lessons From the Forest Sector." *Review of European, Comparative & International Environmental Law* 16, no. 2 (July 2007): 158–72.

Chan, Sewell. "Bloomberg Urges U.N. to Act on Climate Change." *New York Times*, February 11, 2008, https://cityroom.blogs.nytimes.com/2008/02/11/bloomberg-urges-un-to-act-on-climate-change/.

Chudnoff, Martin. *Tropical Timbers of the World*. Madison: US Department of Agriculture, Forest Service, Forest Products Laboratory, 1980.

Clausen, Carol A. "Biodeterioration of Wood." In *Wood Handbook: Wood as an Engineering Material*. Centennial ed. General Technical Report FPL, 14.1–14.6. Madison: US Department of Agriculture, Forest Service, Forest Products Laboratory, 2010.

Corner, James, and Ricardo Scofidio. *Designing the High Line: Gansevoort Street to 30th Street*. New York: Friends of the High Line, 2008.

David, Joshua, and Robert Hammond. *High Line: The Inside Story of New York City's Park in the Sky*. New York: FSG Originals, 2011.

Dunlap, David W. "On West Side, Rail Plan is Up and Walking." *New York Times*, December 22, 2002.

Eguchi, Eduardo. Personal correspondence, May 28, 2012.

Fearnside, Philip M. "The Roles and Movements of Actors in the Deforestation of Brazilian Amazonia." *Ecology and Society* 13, no. 1 (2008).

Ferré-Sadurní, Luis. "Could the Rockaways Survive Another Sandy?" *New York Times*, July 13, 2017, www.nytimes.com/2017/07/13/nyregion/rockaways-beaches-hurricane-sandy.html.

Forest Products Division, US Department of Commerce. *California Redwood and Its Uses Trade Promotion Series No. 171*. Washington, DC: US Government Printing Office, 1937.

Forest Stewardship Council International. "Facts and Figures." https://ic.fsc.org/en/facts-and-figures.

Geist, Helmut J., and Eric F. Lambin. "Proximate Causes and Underlying Driving Forces of Tropical Deforestation." *BioScience* 52, no. 2 (February 2002): 143–50.

Hanna, Ian. Personal correspondence, March 12, 2013.

Harris, Diane. "Writing a Modern Landscape: Thomas Church as Author." *House and Home*, April 1956.

The High Line. Accessed December 18, 2018, www.thehighline.org/design/design-team-selection/field-operations-diller-scofidio-renfro.

Hunt, George M., and George A. Garratt. *Wood Preservation*. New York: McGraw-Hill Book Company Inc., 1953.

Ibanez, Daniel, Jane Hutton, and Kiel Moe, eds. *Wood Urbanism: From the Molecular to the Territorial*. Barcelona: Actar, 2018.

Instituto Floresta Tropical. Accessed November 17, 2018, http://ift.org.br/en/.

Jost, Daniel. "Changing Places: Degrees of Preservation: Three Residential Landscapes by Thomas Church Face Challenges Under New Ownership." *Landscape Architecture Magazine* (January 2009): 30–40.

Keating, Tim. Personal correspondence, February 26, 2013.

Kretschmann, David E. "Mechanical Properties of Wood." In *Wood Handbook: Wood as an Engineering Material*. Centennial ed. General Technical Report FPL. Madison: US Department of Agriculture, Forest Service, Forest Products Laboratory, 2010.

Laird, Sarah A., and Alan Robert Pierce. *Tapping the Green Market: Certification and Management of Non-Timber Forest Products*. London: Earthscan, 2002.

Landscape Architecture Magazine (Advertisements). *Landscape Architecture Magazine* 30 (July 1985).

Landscape Architecture Magazine (Advertisements). *Landscape Architecture Magazine* 80 (July 1990).

Landscape Forms (Advertisement), *Landscape Architecture Magazine*. Volume 64, 1973.

Lentini, Marco. Personal correspondence, May 28, 2012.

Loughran, Kevin. "Parks for Profit: The High Line, Growth Machines, and the Uneven Development of Urban Public Spaces." *City & Community* 13, no. 1 (March 2014): 49–68.

Moynihan, Colin. "Destination New York City." *Metropolis* 19, no. 8 (June 2000): 50.

Nogueira, Marlei. Personal correspondence, May 29, 2012.

NYC Press Release, 2008, PR-045-08. "Mayor Announces Plan to Reduce the Use of Tropical Hardwoods." February 11, 2008.

"A Protest at Great Height." *New York Times*, August 17, 1988.

Rainforests of New York. Accessed February 6, 2018, http://rainforestsofnewyork.net

Reichl, Alexander J. "The High Line and the Ideal of Democratic Public Space." *Urban Geography* 37, no. 6 (2016): 904–25.

Robbins, Liz. "Clash Over Future of Wood From the Storm-Torn Rockaway Boardwalk." *New York Times*, November 15, 2012, www.nytimes.com/2012/11/16/nyregion/rivals-vie-for-wood-from-storm-torn-rockaway-boardwalk.html?_r=0.

San Francisco Environment. "Chapter 8 Tropical Hardwood and Virgin Redwood Ban." Accessed January 20, 2019, https://sfenvironment.org/policy/chapter-8-tropical-hardwood-and-virgin-redwood-ban.

Schulze, Mark, James Grogan, Chris Uhl, Marco Lentini, and Edson Vidal. "Evaluating Ipê (*Tabebuia, Bignoniaceae*) Logging in Amazonia: Sustainable Management or Catalyst for Forest Degradation?" *Biological Conservation* 141, no. 8 (August 1, 2008).

Schulze, Mark, James Grogan, and Edson Vidal, "Forest Certification in Amazonia: Standards Matter." *Oryx* 42, no. 2 (April 1, 2008): 229–39.

Shortle, Walter C., and Kenneth R. Dudzik. "Wood Decay in Living and Dead Trees: A Pictorial Overview." *United States Department of Agriculture, General Technical Report NRS-97.* Pennsylvania: U.S. Forest Service, 2012.

Siegel, William C., and Clark Row. "US Hardwood Imports Grow as World Supplies Expand." *Forest Service Research Paper SO-17.* U.S. Department of Agriculture, 1965.

Smith, Amy. Personal correspondence with author, March 22, 2013.

Smith, Kevin T. "Compartmentalization Today." *Aboricultural Journal* 29 (2006): 173–84.

Soloman, Alan. "It's Tropical on the Highline." *Sawkill Lumber Co.* February 13, 2015. Accessed February 1, 2018, http://sawkil.com/its-tropical-on-the-highline/

Sternfeld, Joel. *Joel Sternfeld: Walking the High Line.* Göttingen: Steidl, 2012.

Teixeira, Marcia. Personal correspondence, May 28, 2012.

Thompson, J. William, and Kim Sorvig. *Sustainable Landscape Construction.* Washington, DC: Island Press, 2000.

Treib, Marc, and Dorothée Imbert. *Garrett Eckbo: Modern Landscapes for Living.* Berkeley: University of California Press, 2005.

Treib, Marc. "Axioms for a Modern Landscape." In *Modern Landscape Architecture: A Critical Review*, edited by Marc Treib. Cambridge, MA: The MIT Press, 1994.

U.S. Consumer Product Safety Commission. "What You Should Know About CCA-Treated Wood." *CPSC Publication 270, 062111.* Accessed November 17, 2018, www.cpsc.gov/PageFiles/122137/270.pdf.

Walker, Timothy. "Slave Labor and Chocolate in Brazil: The Culture of Cacao Plantations in Amazonia and Bahia (17th–19th Centuries)." *Food and Foodways* 15, no. 1–2 (June 6, 2007): 75–106.

Wei, Ruby. Personal correspondence with author, March 5, 2019.

Wellesley, Laura. "Illegal Logging and Related Trade: The Response in Brazil." *A Chatham House Assessment*, 2014. Accessed November 17, 2018, www.chathamhouse.org/sites/default/files/publications/research/20141029IllegalLoggingBrazilWellesleyFinal.pdf.

World Wildlife Fund. "Guide to Lesser Known Tropical Timber Species." Accessed November 18, 2018. www.worldwildlife.org/publications/guide-to-lesser-known-tropical-timber-species.

Zabel, Robert A. *Wood Microbiology: Decay and Its Prevention.* San Diego: Academic Press, 1992.

Zhouri, Andrea. "Global-Local Amazon Politics: Conflicting Paradigms in the Rainforest Campaign." *Theory, Culture & Society* 21, no. 2 (2004): 69–89. doi:10.1177/0263276404042135.

Epilogue

On the Chincha Islands, park rangers look for signs of returning bird populations; on Vinalhaven, deep quarry pools reflect light; the Carrie Furnaces stand awaiting future development; on Rikers Islands, the former tree nursery site lies beneath a bloated incarceration complex; and in northern Brazil, a newly elected government has promised to reduce environmental protections and indigenous power in the Amazon.[1] From these and similar landscapes, fertilizer, stone, steel, trees, and lumber moved out into the world – including to the particular New York City landscapes featured in this book. The flow of these materials played a role in successive waves of urban development: the fertilization of industrial agriculture to feed growing urban populations, the paving of city streets in granite to formalize and enhance the circulation of capital, the restructuring of the ground plane in steel to facilitate vehicular movement out of the city while making a modern public landscape, the expansion of the urban tree canopy as a means to improve the quality of urban living, and the resurfacing of obsolete infrastructure to make new leisure landscapes in tropical hardwoods. These materials participate in the "heroic" narrative of the growth and development of a major metropolis.

We are used to thinking of urban development as if it happens in one place only, as if it were an additive process – a series of successive constructions, improvements, and modernizations. But as this book has explored, each of these developments is materially linked to another place: a largely invisible yet very connected *elsewhere*. The five elsewheres featured in this book are in some ways idiosyncratic. I arrived at these locations by following a lead, starting with one material in a single-designed New York City landscape and ending up wherever that material trajectory took me. I ordered the chapters and selected the New York City landscape materials to highlight changing material practices and urban developments, but any number of other cases could have substituted for them, and each would have drawn a different map of production sites. But, despite the fact that the set of five production landscapes was assembled by some happenstance, a few baseline observations come to light. First, these cases all reveal conditions of labor exploitation and sometimes resistance to it; second, these landscapes are generally characterized by physical or perceived remoteness; and third, the fate of these landscapes and the people who inhabit them have been greatly impacted by the cycle of material boom and bust – of the intensity of material exploitation and also of industry's retreat.

The production of different construction materials requires different types of labor; to cut trees or mine ores makes for very different physical actions, responsibilities, and ambiences. But the cases in this book reveal that whether or not workers are legally protected and empowered has far more to do with the risks, benefits, and conditions in which one labors. The material trajectories reflect a range of these conditions, from slavery to the development of unionized power, across international borders and within them. Three of these five cases – guano, trees, and wood – are directly or tangentially stories of coerced or unpaid laborers: the enslaved Chinese workers captured and incorporated into the "post-abolition" apparatus of guano harvesting; the unpaid incarcerated nursery workers on Rikers Island; and more tangentially, the deforestation of high-value hardwoods in the Amazon basin has been linked to so-called "debt slavery," a modern form of indentured labor. The other two chapters – on granite and steel – highlight worker resistance and organization in the face of egregious labor conditions and injustice. That both the earliest (guano) and most recent cases (wood) in this book are in some ways related to coerced and unjust labor contradicts the common assumption that global working conditions have universally improved over time. While the twentieth century saw substantial improvements in working conditions within more developed nations, the shift to overseas markets effectively outsources not only jobs, but jobs with unsafe and unjust working conditions. Furthermore, gains made in domestic U.S. labor rights have been eviscerated in recent decades as work moves towards the informal economy where the types of security gradually earned by unions have become less and less present.

That three of the five production landscapes from this book are islands – the Chincha Islands, Vinalhaven, and Rikers – is partly a coincidence, but also a reminder of how physical geographies shape material production. Guano accumulated on the Chincha Islands because of a unique biophysical condition, but the remoteness of the location allowed governments invested in the guano trade to turn a blind eye to the horrendous conditions by which it was harvested. Vinalhaven became a key granite source as the island condition facilitated quick and profitable marine transport of the heavy stone from the quarries. And Rikers Island, while so close to Manhattan, was nevertheless a place of "away" – a longstanding repository that kept the city's unwanted citizens, waste, and then incarcerated people at a distance. As discussed in Chapter 5, ipe is harvested from the northern Brazilian Amazon forest, a remote and vast landscape. The Carrie Furnace site near Pittsburgh is the only outlier in this regard. As a production plant rather than a source of raw materials, the Carrie Furnaces integrated multiple materials in the production of pure iron and did so in a dense urban context. The plant is neither remote nor invisible, and this industry has powerfully shaped Pittsburgh's urban development. Even though the earliest case of the book did feature a global-scale material trajectory, in general over the twentieth century we have witnessed material chains getting longer and more complex, their sources ever more distant and unknown.

And finally, by looking at these stories together, we can observe how material production landscapes are impacted by cycles of economic boom and bust. We can see how the exploitation or production of a valuable material commodity can charge a place with activity and employment, as during peak times Vinalhaven's population multiplied, or steelworkers in Pittsburgh aggregated labor power. However, the events that eventually lead to a bust – the overexploitation of a resource like guano, or the transfer of an industry (like steel) overseas – often leave a vacuum, and communities struggle to build new economic opportunities. Also, the extractive or productive activity in question radically changes the land, removing its literal and figurative bedrock and depositing industrial residue in the soil and water, which are then absorbed into the skin and bones of the people living nearby. These landscapes are characterized by the exhaustion of both human and non-human resources, as profits move to the next frontier.

Towards reciprocity

Taken together, patterns of labor injustice and resistance, the remoteness of production sites, and cycles of exhaustion related to material production illustrate Jason E. Moore's notion of how capitalism organizes "cheap natures," rather than a more palatable story of modern progress.[2] As mentioned in the Introduction, the word "reciprocal" as part of this book's title is used in an aspirational sense. The relationships traced between production and consumption landscapes, among various people working along commodity chains, and between people and the land they depend on for survival, are far from reciprocal in the true sense of the word. As entrenched as these unequal relations are, the purpose of this book is, nevertheless, to try to understand and push back at them. Taking for granted that how we think about materials, land, and construction projects frames the way we extract, specify, and build, the following thought experiments mark possible trajectories forward.

First, what if we saw construction materials not as fixed things but rather as physical continuities of matter, connected to land and people?[3] What if, like x-ray vision, we could see materials beyond their commodity status, as things that have values far greater and more complex than their market price? What if we saw urban development not as a story of human ingenuity on a green backdrop, but instead as a co-production with other species, materials, and distant landscapes? If we could unsee or unlearn the pervasive idea that materials are inert, exist in a single state, and are subservient to human need alone, we could instead grasp materials' agencies and observe more clearly the flows and interdependencies between construction and the more-than-human world.

Second, what if we conceptualized material production as part of the project, rather than external to it? While contemporary landscape architecture has expanded the scope and scale of practice and research to address the world's most pressing challenges, acknowledging material production as integral to design projects would address larger scales and concerns without necessarily changing the footprint of the work site. Here we can take inspiration from the food justice movement, whose insistence on articulating the links between land practices, ecological dynamics, and social justice has drastically changed the discussion about where food comes from. By challenging the normative models of industrial agriculture, by reclaiming heritage species that were once central to regional cuisine, and by supporting cooperative and community-based projects, the meal itself is different. While construction materials are very different from agricultural products, looking at how these connections have become common-place is useful. As with food, we need better fluency about the relationships of landscape-making; only by understanding these relationships can we begin to design in ways that might change them – to engage with workers, other species, and landscapes "elsewhere." Through a close reading of the relationships and people associated, we can design in ways that strengthen relationships, share benefits, and take responsibility; rather than outsource risk, we can support socially just and sustainable land-management relationships elsewhere.

And third, how can the act of design itself – the particular forms and configurations in which materials are assembled, plants are established, public activities are facilitated – actually address some of the larger material relationships at play in the materials used to construct it? There is a longstanding tradition of didactic landscape architecture projects intending to teach people something through references and symbolic approaches. But perhaps designers could assemble materials to actually communicate more about them – about their source landscape, their formation, the labor behind them? If at the core of current ecological crises lies a profound separation from land, from the consequences and interdependencies that humans share with the land, how can material design engage this? The textures, smells, structures of particular materials give people tactile and intimate contact with fragments of distant landscapes and their myriad social and ecological relationships. As the stories in this book have shown, designed public landscapes always reflect contemporary ideologies about human relationships to nature; they offer new forms of thinking and engaging with materials. As sites of intense public activity, designed public landscapes have the potential to capture the popular imagination, spark debates, and advocate for change about the issues that matter most. Perhaps it is through the materials that we come into closest contact with that we can connect with the most distant.

Notes

1 Dom Phillips, "Jair Bolsonaro Launches Assault on Amazon Rainforest Protections," *The Guardian*, January 2, 2019, www.theguardian.com/world/2019/jan/02/brazil-jair-bolsonaro-amazon-rainforest-protections.
2 Jason W. Moore, *Capitalism in the Web of Life: Ecology and the Accumulation of Capital* (New York: Verso, 2015), 291–305.
3 Recognizing the fallacy of universal "we" when it comes to responsibility for environmental and social injustices, the use of "we" here refers to individuals involved in the design of the built environment.

Bibliography

Moore, Jason W. *Capitalism in the Web of Life: Ecology and the Accumulation of Capital*. New York: Verso, 2015.

Phillips, Dom. "Jair Bolsonaro Launches Assault on Amazon Rainforest Protections." *The Guardian*, January 2, 2019, www.theguardian.com/world/2019/jan/02/brazil-jair-bolsonaro-amazon-rainforest-protections.

Acknowledgments

Just like the materials followed in this book, my gratitude is far flung.

It has been a pleasure and an education to spend time with people living in, working on, and struggling for the landscapes described in this book. So much of the content of this book comes from conversations had while walking, driving, boat riding, or corresponding with these people: In Paracas, Peru, Jhuneor Paitan Ñahui, Rodrigo Ramirez, Mariano Dalverde, and Jorge Tarazona Paredes of SERNANP introduced me to guano on the Chincha Islands and their organization's efforts to sustain its vital ecosystem; in Vinalhaven, Maine, Bill Chilles toured me around the granite landscape that he knows so well; in Pittsburgh, Ron Baraff from Rivers of Steel brought the quiet Carrie Furnaces to life, Bob Bingham narrated Nine Mile Run, while Joel Tarr, Ken Kobus, and Erin Deasy shared their knowledge about the steel landscape; in Brazil, Brian Holz, Marcia Teixeira, and Eduardo Eguchi generously offered their time to explain lumber movement through Belém, and Marco Lentini and Marlei Nogueira of the Instituto Floresta Tropical hosted me at the Roberto Bauch Forest Management Center in Paragominas; in and around New York City, Tim Keating shared stories and challenges of decades of activism against tropical deforestation, Marie Warsh of the Central Park Conservancy introduced me to future and past park constructions, Robert Zappala shared memories of the Parks Department Nursery on Rikers Island, Angelyn Chandler and Ruby Wei of the Parks Department offered contacts and expertise, and former Parks Department Director of Street Tree Planting Matthew Stephens and nursery owner Joe Sipala toured me around rows of future New York City street trees in Long Island.

Archival photographs and documents are an important part of this research, and I am grateful to the many archivists, librarians, and collections that have shared their expertise and resources, including Rebekah Burgess at the New York City Parks Photography Archives; Elizabeth Bunker, Bill Chilles, and Sue Radley at the Vinalhaven Historical Society; the New York Historical Society; Steven Rizick at the New York City Parks Olmsted Center Map File Room; the New York Botanical Garden; the New York City Municipal Archives; the Museum of the City of New York; the New York Public Library; the University of Pittsburgh Archives and Special Collections; the University of Minnesota Duluth Archives and Special Collections; the

Carnegie Museum of Art; the Smithsonian Institution Archives of American Art; the Harvard Business School; the California Redwood Association; the New Bedford Whaling Museum; David Rumsey Map Collection; the Forest Products Laboratory; and the United States Geological Survey. Several photographers, artists, landscape architects, and activists have generously shared photographs and projects for this book: I would like to thank Rick Darke, Jonathan Ellgen, Cynthia Goodson, Jill Hubley, Roman Iakoubtchik, Margaret Morton, Tomas Muñita, Michael Van Valkenburgh, and Rainforest Relief for contributing their work.

The majority of this book was researched and written while I was working at the Harvard Graduate School of Design, and I am truly grateful to Charles Waldheim for offering the space and encouragement to initiate and sustain this research. There, a Dean's Junior Faculty Grant funded travel that formed the basis of the research, and Gary Hilderbrand, Anita Berrizbeitia, and Niall Kirkwood provided support in crucial ways. I'm grateful for wonderful colleagues who offered insights and support during that time, and I'd especially like to thank Emily Waugh, Jane Wolff, and Sonja Duempelmann for inspiring me to think about how stories are told, Jill Desimini for reading early drafts, and Kiel Moe for collaborations that fed the project in different ways. The final segment of the book was completed at the University of Waterloo School of Architecture, and I am thankful for the warm welcome from faculty, staff, and students, and especially Anne Bordeleau. I've had the pleasure of working with and learning from students at both schools in courses related to the theme of this book and as research assistants. I'd especially like to thank those that contributed to this book: Senta Burton, Dane Carlson, Manuel Cólon-Amador, Cathy De Almeida, Emily Drury, John Evans, Safira Lakhani, Jennifer Lee Mills, Sophie McGuire, Kelly Murphy, Anne Webster, Sarah Michele Richmond, Sneha Sumanth, and Yi Ming Wu.

Multiple editors have helped shape early drafts of articles that became significant portions of this book: thanks to Kelly Shannon and peer reviewers at the *Journal of Landscape Architecture* for the article "Reciprocal Landscapes: Material Portraits in New York City and Elsewhere"; Amy Kulper and the *Journal of Architectural Education* for "On Fertility"; Leah Whitman Salk, Jennifer Sigler, and Megan Sandberg at *Harvard Design Magazine* for "A Range of Motions"; Daniel Jost of *Landscape Architecture Magazine* for "A Trail of Stumps"; and Claire Lubell at the Canadian Centre for Architecture for "Inexhaustible Terrain." This project began at a symposium at University College London, organized by Ed Wall, Tim Waterman, Douglas Spencer, and Murray Fraser; later Ed and Tim edited a volume based on this event, in which "Reciprocal Landscapes: Material Portraits" was reprinted. Thanks to all of these publications for permissions to draw from and reproduce portions of these texts. I am grateful to Alpa Nawre, Ed Wall, Kim Förster, Melissa Cate Christ, Jonathan Solomon, Alissa North, Jacob Mans, and Amy Kulper for inviting me to share this work while it was still in progress. A MacDowell Colony Fellowship gave me the opportunity to assemble archival images and finalize the manuscript in a sublime environment alongside inspiring people.

I'm grateful to many friends and family members for long-standing exchanges on landscapes, life, and matter that have inspired this project in so many ways. Jeffrey Malecki edited the text and provided crucial feedback throughout, and I am grateful for his steady, insightful eyes. Towards the end, Sameer Farooq stepped in with important visual insights and advice, thank you. At Routledge, I would like to thank the peer reviewers who provided constructive comments in earlier iterations, and I am extremely grateful to Grace Harrison, Emily Collyer, Aoife McGrath, Louise Baird-Smith, and the publication team for their encouragement, responsiveness, and patient guidance. Most of all, to Adrian, thank you for everything.

Index